Understanding Digital Electronics

Written by: Gene McWhorter
Longview, Texas
Staff Consultant, Texas Instruments Information
Publishing Center

Managing Editor: Gerald Luecke, MSEE
Mgr. Technical Products Development
Texas Instruments Information Publishing Center

Foreword is by Jack S. Kilby who, while in the employ of Texas Instruments Incorporated, invented the integrated circuit in 1958 and is co-inventor of the first handheld calculator. Mr. Kilby was presented the National Medal of Science by the President of the United States in 1970 as recognition of the integrated circuit invention.

TEXAS INSTRUMENTS

P.O. BOX 225012, MS-54 · DALLAS, TEXAS 75265

This book was developed by:

The Staff of the Texas Instruments Information Publishing Center
P.O. Box 225012, MS-54
Dallas, Texas 75265

With contributions by:

Jim Allen   Gene Marcum
Les Mansir   Frank Walters
Gerald Luecke   Ralph Oliva

Appreciation is expressed to Walt Matzen, Phil Miller, Marcus Allen, Kirk Allen, and Doug Luecke for their valuable comments.

Artwork and layout by: *Word Processing:*

Plunk Design   Betty Brown

ISBN 0-672-27013-7
Library of Congress Catalog Number: 84-51467

Second Edition
Second Printing

About the cover:

Represented on the cover are the components of a digital system that fits in your hand — a calculator: silicon slices in various stages of completion that yield the integrated circuits that are the heart of the system, a liquid crystal display, the interconnecting substrate, and a finished calculator.

Table of Contents

Foreword

Since 1958, in a period of about thirty years, digital electronics has become one of the most rapidly growing industries in the world. In that year, most of the digital applications were found in computers, and probably less than 1200 machines were completed during the year. Today, one manufacturer, of several, builds more than 1000 calculators every hour of the working day. To emphasize further the advances accomplished in such a short period of time, some of these new calculators have more computational capability than the 1958 computers. Quite a digital evolution!

Now, calculators and computers represent only one of the many uses of digital electronics. Old familiar analog circuits in consumer products such as radios and televisions have been replaced with circuits using digital techniques. New electronic products for new markets such as microwave ovens, sewing machines, TV games, are springing up each year. In fact, digital circuits are even replacing mechanical parts like gears and pinions — as in the modern digital electronic watches.

This rapid growth has come about because of the almost ideal match between the digital electronic requirements and the capabilities of the integrated circuit. Digital circuits give "off"-"on" answers, permitting the use of components with wide tolerance which are easier to make. Because they are handling only information, they can operate with very low power. As a result, they can be very small physically and many thousands of digital functions can be built on a single integrated circuit chip at very low cost.

It is also the very low cost which has been responsible for the rapid growth in digital functions. A digital circuit that makes a decision — called a "gate" — which cost several dollars in 1958 can be obtained as a part of an integrated circuit today for less than a tenth of a cent. A reduction of more than a thousand to one! These decreases in circuit costs are continuing, helping digital electronic systems of the future to cost even less — and to find even wider uses.

The technical and economic forces which caused this rapid growth of digital techniques will open up new applications areas for electronics. We are truly on the threshold of an era where digital electronics will have a pervasive presence.

Jack Kilby

UNDERSTANDING DIGITAL ELECTRONICS

Preface

If you have a junior-high-school background in electricity, plus a curiosity about how things work and a general awareness of electronics in use all around you, this book is for you. It's for you whether you are a PhD who hasn't studied digital electronics yet or an eighth-grader who wants to build his own digital computer.

This book won't show you how to build that computer. It will do something more important. It will give you *understanding*— understanding of the electronic circuitry in many types of digital electronics, from the basic idea of a transistor circuit saying "yes" or "no," to entire digital systems made up of thousands of such circuits. This understanding will serve you well whether you have your hands into real hardware or simply wish to be in touch with the most revolutionary technology of our time.

This book is different from many others, in that it's a *self-teaching course*. That means it builds your understanding step by step. You shouldn't skip around in the book or try to pick out individual things to learn. Read one chapter at a time, beginning with Chapter 1. Quizzes are provided at the end of each chapter to review main points learned in the chapter.

Try to master each chapter before you go to the next. This is to make sure you have a solid background for learning more advanced things later on. Each chapter will move you rapidly to a new level of understanding.

A glossary for all special words and an index are provided to aid in the understanding and use of the material.

We who have prepared this book hope that, as you go through it, you will feel some of the excitement that comes from learning about the marvelous things that digital circuitry can do—and the even more marvelous things that are yet to come from this fascinating new creation.

Let's Look at a System

Stop! Think a minute! Haven't you been curious about those electronic games that you play on a television screen? Have you ever wondered how an electronic digital watch works, or a hand-held calculator? How about the computerized control systems used in automobiles, or that computer used at your bank, or the office, or in a small business, or for credit cards?

All of these systems are digital electronic systems. "Digital electronics" means the kind of electrical circuitry found in such systems.

This kind of circuitry is very different in design from that found in older, more common electronic systems such as radio and television receivers, high-fidelity sound recording and playback systems, and electric guitars. These systems use another style of electrical circuit design called linear or "analog" electronics.

What's special about digital electronics?

Digital systems, which use electricity in an entirely different way than the older, more common analog systems, are providing systems with more capability, in less space, at less cost.

Digital and analog systems are similar in that they both use electricity, electronic devices such as transistors and diodes, and various other electronic parts. You can't always tell by looking inside a system whether it's digital or analog. The difference is in the way the systems *use* electricity — the things they make electricity do. This different way of dealing with electricity gives digital systems the ability to do almost unbelievably complicated things for you, without being very big or costing a lot of money.

It's timely and important to learn about digital electronics because these sophisticated, compact, and economical systems are getting even *more so* as time goes on. They are cropping up in more and more places — both as replacements for analog systems and as entirely new ideas that were never possible before.

And so to keep up with progress, it's not enough to know about microphones, loudspeakers, transformers, potentiometers, amplifiers, oscillators, mixers, tuners, detectors, filters, waveforms, impedance matching, feedback, frequency response, and other terms common to analog electronics. The wave of the future is with *digital* electronics, including terms such as gates, flip-flops, counters, registers, decoders, binary numbers, TTL, MOS, and microprocessors.

How will this book help you?

In this book, we're going to survey the field of digital electronics, from a light switch in your house (yes, it's a *digital* device!) to a large digital computer. We'll learn the features common to digital systems, and how digital electronics works in a wide variety of applications. We'll see *why* digital methods are revolutionizing the field of electronics. And more than that, we'll learn what to expect in the future from this amazing technology.

Furthermore, we're going to do all this without getting you bogged down in the fine details of circuit design. Because one of the most marvellous things about digital electronics is that you can have a deep, sophisticated understanding of it *without knowing very much about electricity!*

Even if you already know enough about both electricity and digital applications to tinker around a little bit with digital circuits – chances are you'll find in this book a deeper, richer understanding of the subject plus its implications for the present and the future.

What's a familiar digital system?

A hand-held calculator, a sophisticated digital electronic system of small size and low cost, will be used to understand digital electronics without knowing a lot about electricity.

Right away, we're going to find out just what a digital system is, and start learning how digital systems work. Let's begin with the digital system you're probably most familiar with personally – a small electronic calculator, such as the Texas Instruments calculator shown in *Figure 1-1.*If you've got a hand-held calculator or a small desk-top model, stop reading and get it now. If you don't have one, perhaps you can borrow one – or you may want to buy one. It may help your learning and appreciation of this subject a great deal.

Okay, now look at the calculator, and think for a moment about how small and inexpensive it is, considering the amazing things you know it can do. Just a few short years ago, an electronic calculator that could add, subtract, multiply, and divide was as big as a large electric typewriter and cost maybe five hundred dollars. And this illustrates what we said earlier about digital systems getting more sophisticated, smaller, and lower in cost as time goes on.

**Figure 1-1.
A Hand-held Digital
System**

UNDERSTANDING DIGITAL ELECTRONICS

Now let's consider what this calculator can do. Turn it on. Press the "3" key, noting what happens. The result may not seem very impressive at first – just a matter of a number "3" appearing in the display, right?

What goes on inside a calculator?

The simple function of adding two numbers requires several complex operations to be performed electronically by the calculator: transmitting, adding, storing and displaying.

But ask yourself what made this "3" appear. Look closely at the lighted number itself. If your calculator is like most, the "3" consists of five small lighted segments, out of seven segments that can be lighted. When all seven segments are lighted, you get an "8". The segments in your display may be tiny red bars, rows of even smaller red dots, larger green bars, or dark bars not illuminated. These are all different ways to make the same basic pattern of seven segments.

Now consider what made the particular five segments turn on to form the "3". Apparently, pressing the "3" key *sent some information* somewhere inside the calculator – some information saying, "Remember number 3." And somewhere inside, something is *remembering* "3". And somehow this remembered "3" is making five particular segments of the display *light up.*

Now go through the steps for adding five to the three and getting the total. The particular keys you press at which times depends on just what kind of calculator you have. Most likely, you press the "plus" key, then the "5" key, and finally the "equals" key. Note what happens as you go through the necessary steps to get the total of five and three.

First, the "3" was replaced by a "5," right? So where did the "3" go? Apparently, it was still being remembered somehow without being displayed. And the "5" was lighted up in the same way the "3" was earlier. If you pressed "plus" before the "5," then something inside had to remember you wanted to add the next number. If you pressed "equals" after the "5," this apparently caused the two numbers to be added, because now an "8" is being displayed. But what inside the calculator *figured out* this answer? And what happened to the 5 and the 3? Where are they now?

Obviously, there are some pretty complicated things going on inside this machine, even when we simply add two numbers less than ten. When we have answered the questions as to how these numbers were *transmitted* from the keyboard, how they were *added,* how they were *stored,* and how they were formed on the *display,* we will have answered the questions of what a digital system is, and how it works. So let's get started.

How can we simplify a calculator for study?

The functions of a calculator can be clearly and easily illustrated by using a symbolic diagram called a "schematic" diagram.

Let's limit our discussion at this point to a very simple imaginary calculator — one that will only add, subtract, multiply, and divide. Its display will handle numbers with only eight digits (numerals). It won't work *exactly* the same way as the calculator in your hand, so you can put yours down. But keep it handy for reference.

Furthermore, let's say that the electronic circuitry in our example calculator is the simplest possible to handle these limited tasks. But the general way the circuitry works is very much like the operation of most real calculators that will do more sophisticated things.

First, let's consider the main parts of the machine, as shown schematically in *Figure 1-2*. (A "schematic" drawing of a circuit is one using simple *symbols* for the various parts and the interconnecting wires.) The large block at the bottom represents an integrated circuit — words we'll abbreviate to "IC." The 22 arrows pointing in and out of the IC represent wires, and the arrowheads indicate the direction electric current flows.

**Figure 1-2.
Schematic Diagram of
Example Calculator**

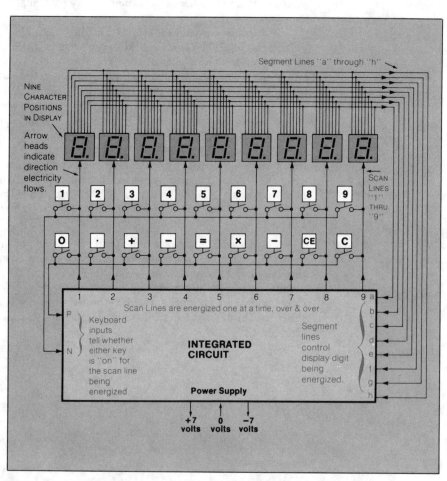

What's an integrated circuit?

The digital integrated circuit is a complete electronic circuit composed of thousands of microscopic semiconductor components all of which operate in a digital mode.

Integrated circuits are the main reason digital systems are becoming more and more sophisticated, compact, and economical. They are a method of mass-producing complicated electronic circuits containing thousands of transistors, diodes, resistors, capacitors and the interconnecting wires in an unbelievably tiny form.

Figure 1-3 shows a Texas Instruments calculator integrated circuit. It's a little package about an inch and a half long, half an inch wide, and an eighth of an inch thick (38 by 13 by 3 mm), with metal strips (pins) for electrical terminals. These strips are connected on the inside to a little "chip" of semiconductor material called silicon. The chip, which is about a quarter of an inch square (6mm) and not much thicker than the pages of this book, is shown in the enlarged photograph of *Figure 1-3*. As you can see, there are so many transistors interconnected with other components on this chip, packed so close together, that you can't tell them apart. Many small calculators have *all* their electronic circuitry packed into just one integrated circuit (not counting the batteries, keyboard, and display).

That's a brief look at ICs. We'll explain them further in a later chapter. But for now, let's move on with our discussion of the calculator.

What are the calculator parts outside the IC?

Calculator keys provide inputs to the system by closing a switch when a key is pressed that connects a keyboard input line to a scan line.

Looking back at *Figure 1-2*, we see 18 little blocks representing the calculator keys. Under each key is a schematic symbol representing a switch. One pole (terminal or connection) of each switch is connected to a horizontal "keyboard input" wire labeled N or P. The other pole is connected to a vertical "scan line" wire (numbered 1 through 9). Pressing a key closes (turns on) a switch for a moment. This allows electric current to flow from one of the vertical scan lines to one of the horizontal keyboard-input lines

Notice that a custom in schematic diagrams is to use a little *black dot to show when two wires are connected.* If two lines representing wires cross *without* a dot, they're not connected. Many of these "wires" would actually be little metal strips on a printed-wiring card. Real wires (or strips) are not always laid out so straight and neat as they appear in a schematic diagram.

Above the keys are nine somewhat larger blocks called "character positions." These blocks form the display, where numbers as long as eight digits can be shown, in addition to a minus sign and various symbols for errors. We'll get to these in a moment. But first, let's talk about how the keys transmit numbers and commands to the IC chip.

**Figure 1-3.
A Typical Calculator IC
Chip**

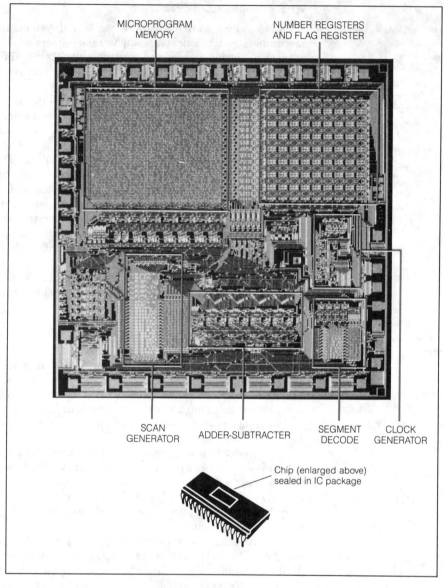

MICROPROGRAM
MEMORY

NUMBER REGISTERS
AND FLAG REGISTER

SCAN
GENERATOR   ADDER-SUBTRACTER   SEGMENT
DECODE

CLOCK
GENERATOR

Chip (enlarged above)
sealed in IC package

How do numbers get inside the calculator?

When a key is pressed, pulses from an appropriate scan line are sent on the N or P keyboard input to the integrated circuit. The combined information identifies the key that was pressed.

Figure 1-4 shows a close-up view of part of the keyboard for discussion purposes. At all times, the IC supplies power to one of the nine vertical "scan lines" at a time, over and over, 1 through 9, thousands of times each second. When the IC is ready for the next keystroke, it looks for a signal coming in on the two "keyboard input" lines, labeled "N" and "P." When you pressed the "3" key, the corresponding switch stayed closed long enough for all the scan lines to be energized several times in a row — no matter how quickly you released the key. (Compared to digital circuitry, the fastest mechanical switch is as slow as molasses in January!) And so pulses began arriving at the "N" input line whenever scan line 3 was supplied power in pulses. These pulses coming at these particular times told the IC that the "3" key was pressed.

Similarly, when the "plus" key was pressed later on, pulses began coming in on the "P" line whenever scan line number 3 was energized. And pressing the "5" key caused pulses on input "N" when scan line 5 was energized.

**Figure 1-4.
Inputs From Keys Into
the IC Chip**

The scan lines are energized sequentially by the IC chip.

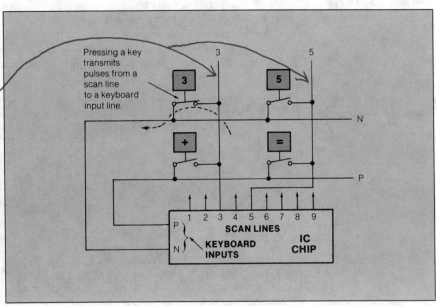

Pressing a key transmits pulses from a scan line to a keyboard input line.

How are numbers shown in the display?

Energized light-emitting diodes (LED's) display the visible numbers, signs, or symbols in the digit positions. Seven individual LED segments make up the characters. An additional one provides a decimal point.

So that's how information gets into the IC from the keyboard. Now let's talk about how numbers are illuminated in the display. Looking at *Figure 1-2* again, each of the nine character blocks is a position for one "character," meaning a numeral digit, minus sign, or error symbol — perhaps including a decimal point to the right. Each of the nine positions is connected to one of the vertical scan lines, and also to eight "segment lines" labelled "a" through "h." Each segment line is connected to all nine character positions and to the IC.

Now look at the detailed view of one of the character positions, shown in *Figure 1-5*. There are seven little light-emitting diodes or "LEDs" forming a figure-8 pattern, and an eighth LED off to the right for a decimal point. ("LED" is pronounced by saying the letters: "L-E-D.") The LEDs are labeled "a" through "h," to match the segment-line designations. These devices are made of a special kind of semiconductor material that gives off light when electric current is passed through them in the right direction.

In addition to the scan line, each LED segment is connected to the IC by a segment line. The scan line, which is connected to all segments, supplies the power to illuminate the LED segments. The proper segments to form the character to be displayed are selected by the IC.

Each LED has two electrical terminals. One terminal on each LED is connected to the scan line coming up to that character position from below (bolder lines), and the other terminal is connected to one of the eight segment lines (lighter lines). To illuminate one LED segment, both its scan line and its segment line must be turned on by the IC, so that current can be supplied by the scan line and returned to the IC by the segment line. (When a *scan* line is "on," it *supplies* electricity. But when a *segment* line is "on," it *accepts* electricity or "sinks" electric current.)

As a result of this arrangement, each character position can be illuminated only when that particular scan line is supplying electricity. And the character (the number or symbol, etc.) that appears at that position is *defined* by which segment lines are turned on to allow current to flow. The IC is able to change the combination of active segment lines every time it energizes another scan line.

Figure 1-5.
Details of Connections in Calculator Display
(*Arrowheads show direction electricity flows*)

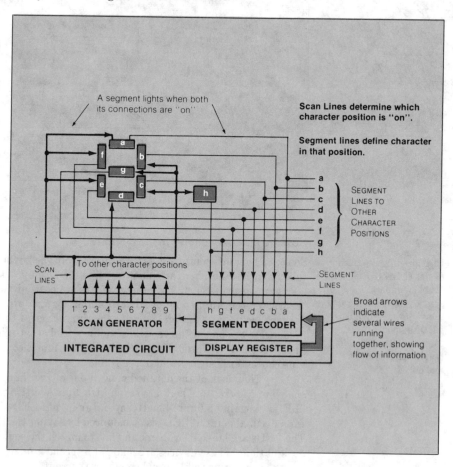

The character is illuminated only as long as its scan line has power input from the IC, which is about one-ninth of the time. However, the scan is so fast, the character appears on all the time.

For example, when scan line 9 and segment lines a, b, c, d, g, and h are "on," a "3" followed by a decimal point appears in the far right position. (Verify this by noting which LEDs in *Figure 1-5* have these labels and then look again at *Figure 1-2.*) Then as scan line 9 goes off and scan line 1 comes on in the regular scan-line sequence, the "3" and decimal point blink off. And the character intended for the far left-hand position blinks on — if any is called for. The blinking is so fast that even though *each character position is "on" only one-ninth of the time*, your eye sees only a steady display.

As you can tell, the IC is working like a demon, even when it's not calculating but merely showing you a number in the display. A thousand times every second, it has to be prepared to switch on a different pattern of segment lines, while watching for pulses on the keyboard input lines. This switching may seem fast to you, but it's actually *slow* compared to many other digital systems that we will discuss in due time.

Because it's capable of scanning with such rapid action, the IC can handle 18 switches and 72 LEDs with only 19 connections. A separate connection for each switch and LED would cost much more (as we will see in a later chapter when we discuss how ICs are made). It would thus cause the calculator to cost much more.

What's inside the integrated circuit?

So that's how information gets into and out of the integrated-circuit chip. (Remember, we called it a "chip" because it's only about ¼ inch on a side and paper-thin.) To know the rest of the story, we've got to look inside this IC.

The subsystems are connected by groups of wires, called buses, that act as a single path way for transmission of data.

·*Figure 1-6* is a simplified diagram showing the main electronic subsystems in the chip simply as blocks. (A "subsystem" is just a smaller system inside a larger one.) The broad arrows represent pathways for information between subsystems. Each of these pathways is really several wires running together to carry simultaneous electric signals. To appreciate how greatly simplified this diagram is, look closely at *Figure 1-3* again. The long, narrow, light-colored strips are thin ribbons of metal acting as wires in the pathways we're speaking of.

Also shown in *Figure 1-6* above are blocks representing the keyboard and display. Let's follow the action as we add 3 and 5.

Figure 1-6.
Major Subsystems in the
Example Calculator

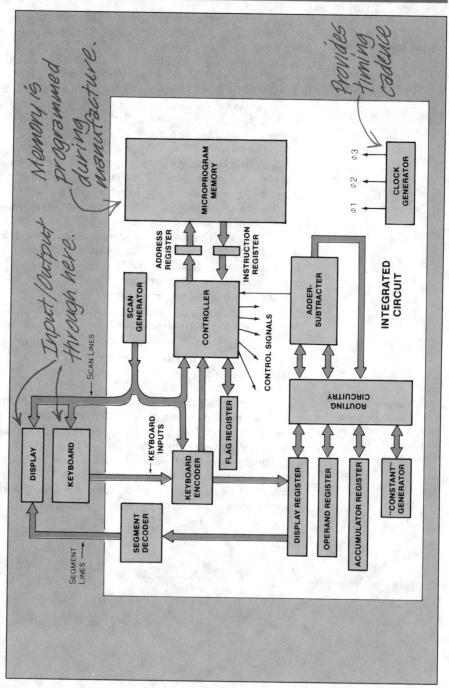

How are the subsystems controlled?

Sets of instructions permanently stored in the microprogram memory during its manufacturing are the basis on which the controller acts to tell each subsystem what to do and when to do it.

The first thing to understand is that all the subsystems are directly linked to the "controller" subsystem by a network of electrical conductors that are not fully shown in *Figure 1-6*. The job of the controller is to tell each subsystem when to act, and what to do. And the controller, in turn, acts merely as an interpreter of instructions that it draws one at a time from a place where they were stored when the chip was made — a place called the "microprogram memory."

**Figure 1-7.
The Address and
Instruction Registers**

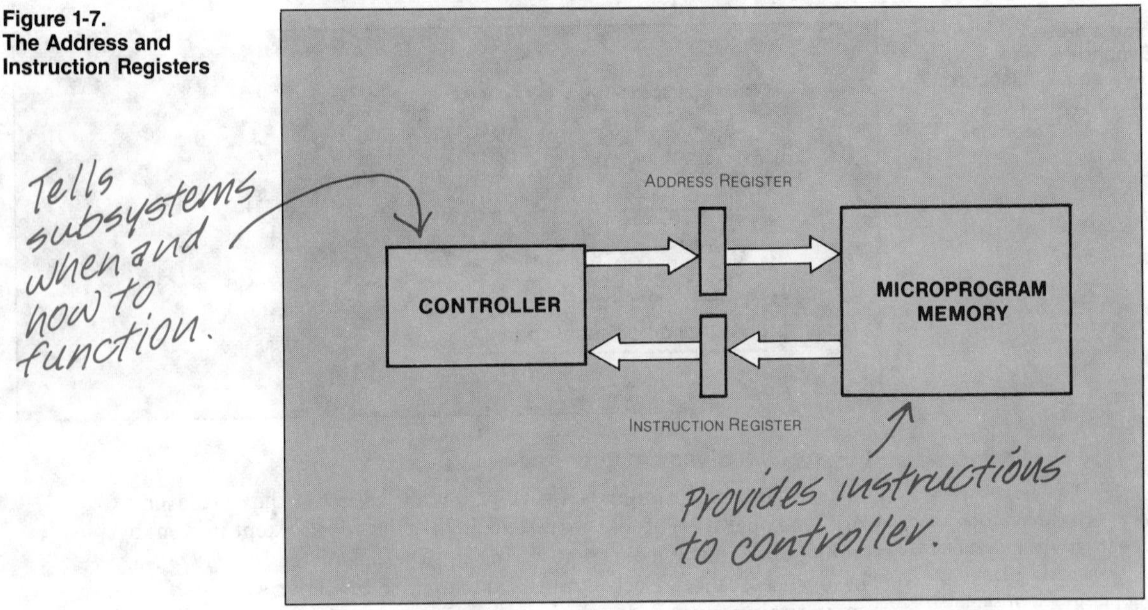

Tells subsystems when and how to function.

ADDRESS REGISTER

CONTROLLER

MICROPROGRAM MEMORY

INSTRUCTION REGISTER

Provides instructions to controller.

The location of an instruction in memory is called the address. The controller obtains an instruction by sending an address to memory. The memory locates the instruction and places it in an instruction register in the controller.

As indicated in *Figure 1-7*, each instruction is permanently stored at a particular location in the microprogram memory. Each storage location and the instruction inside is identified by a number called its "address" — like your house number or apartment number. The controller gets each instruction it needs by putting the correct address number in a temporary storage unit called the "address register." In response, the microprogram memory unit automatically goes to that address, finds the instruction, and immediately delivers a copy of it into another temporary storage unit called the "instruction register," for use by the controller. (As you can tell, a "register" is a storage unit used to hold information for a short while until the information is needed.)

Each instruction is executed for one instruction cycle, the time that each scan line is energized.

Each instruction that the controller looks up this way governs its actions for a period of time called one "instruction cycle." An instruction cycle corresponds to the time during which one scan line is energized – about 100 microseconds (100 millionths of a second). At the end of each instruction cycle, the controller draws (Fetches) another instruction for the next cycle – stepping along one instruction at a time every cycle. If the current instruction doesn't tell the controller how to decide which instruction to use next, the controller automatically picks the one at the next address in sequence in the microprogram memory.

**Figure 1-8.
Synchronizing
Sequential Pulses from
Clock Generator**

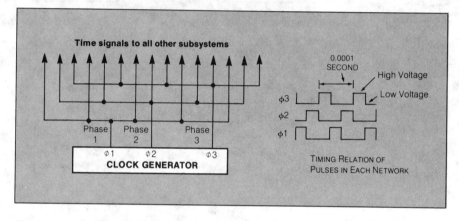

How are operations synchronized?

The clock generator performs synchronization for all subsystem operations by transmitting three timing pulses (clock signals) throughout the system.

Obviously, timing is a very important aspect of the calculator's work. The operation of all subsystems is "synchronized" (kept in step) by timing pulses in three different wiring networks depicted in *Figure 1-8.* These pulses, called "clock signals," are supplied to all parts of the IC from a main timing subsystem called the "clock generator." The three networks, and the pulses each one carries, are called "phase one, phase two, and phase three." The three pulses occur one after the other in a regular cadence, like an orchestra director calling out musical measures in waltz time: "One-two-three, one-two-three." Certain parts of the system will not go into action until they receive these phased timing signals.

What happens before we begin a problem?

With this background information, let's proceed now to add 3 and 5. When we first turn on the calculator, the controller automatically draws (Fetches) instruction number "zero," through the steps we discussed with respect to *Figure 1-7.* This instruction tells the controller to clear out all information in the "register" subsystem as shown in *Figure 1-9.* These registers, as we said earlier, are temporary storage places for numbers and other information. This "clearing" step wipes out any random, meaningless information that may pop up in these registers when the system is first turned on. It's all done in one instruction cycle, by means of a control signal to all registers. (Remember, an instruction cycle takes only 100 microseconds—a tenth of a thousandth of a second!)

Figure 1-9.
Initial Clearing of
Storage Registers by
Controller

The registers
are cleared
to start at
set initial
conditions.

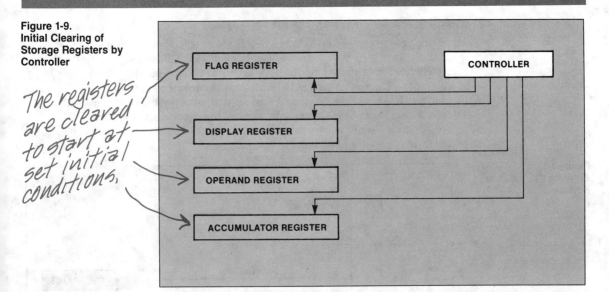

When the calculator's system is cleared, the controller continually scans for signals from the keyboard until one is received. Then it goes to the next instruction.

Other subsystems function simultaneously while the controller is scanning for keyboard signals, such as coordinating clock signals or activating the required segment lines for the characters that are being displayed.

In the next instruction cycle, automatically stepping to the next instruction in the microprogram memory (number 1), the controller is told (all in digital code, of course): "Check for signals from the keyboard. If you *don't* see any signal, follow the same instruction again. But if you *do* see a signal, go on to the next instruction." And so until a keyboard input signal arrives, the controller doggedly sticks to this "keyboard-checking" instruction for cycle after cycle, in time with the beats of the clock signals. The controller is in what we call the "idle routine." All step-by-step sequences of instructions that it follows in doing various tasks end in this routine.

Meantime, while the controller is busy doing this, the "scan generator" subsystem (shown back in *Figures 1-5* and *1-6*) is humming along all by itself, paying no attention to the controller. It's busy counting clock signals, and turning on one scan line after another (as we have already discussed) at the beginning of each 100-microsecond instruction cycle.

And over to the right in *Figure 1-5*, the "segment decoder" subsystem is doing its own thing, too. Its job is to keep the display illuminated with the number digits that are presently stored in the "display register" subsystem, by turning on the appropriate segment lines to receive current at the right times. (The display register is a temporary storage place for an 8-digit number, complete with decimal point and minus sign, if any.) Every time a new scan line comes on, the decoder looks at the next digit position in the display register, and figures out which segment lines to turn on to show this digit in its position in the display. It automatically leaves out any zeroes at the beginning of any stored number, except that it does show you one zero and a decimal point if the register has no number in it—that is, if the register is empty. So that's what it's showing now as we begin to add 3 and 5.

What happens when we press the "3" key?

So—we press the "3" key. (See *Figure 1-10*). Nothing happens until the scan generator turns on scan line number 3. Then a signal is transmitted in keyboard input line N to the "keyboard encoder" subsystem. Knowing which scan line is on, the encoder generates a number "3"—not in the way you would write it on paper but in a special code so that it can be electronically transmitted to the display register and stored there. Recall that we talked earlier about "remembering a 3"—well, the display register is what does that.

**Figure 1-10.
Entering "3" from the
Keyboard**

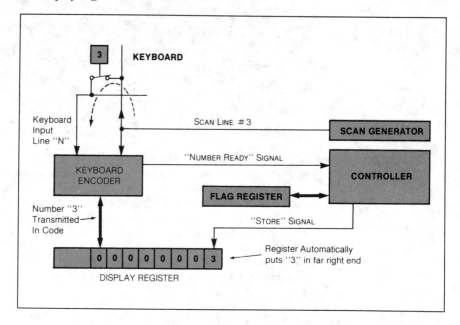

The encoder also sends a signal to the controller, telling it that a number key has been pressed. The encoder doesn't say *which* number key, because the controller doesn't really need to know.

It's not too important right now—but in order to make sure that a key has really been pressed and that the encoder wasn't just picking up some unwanted signal (called "noise"), the controller fetches and obeys some instructions that make it check several times to make sure that a key *was* pressed.

Finding the "3" signal still there, (remember, the switch is slow compared to this digital system!) the controller next has to decide what to do with it. Still following the steps of the programmed routines, it looks for any notes it has previously stored in the "flag register" subsystem shown in *Figure 1-10*—notes with regard to what kind of problem it's doing, and what steps in the problem have already been completed. The flag register is provided for just this purpose—it stores incidental notes, as the program steps are executed, that the controller will need in the future as it completes all of the steps of the problem.

As the controller executes the instructions from inputs received, it continually checks the flag register to determine the type problem it's doing and how far the task is completed.

No notes are found in the flag register, so the controller writes a note in the flag register to remind itself that the first key of a new problem has been pressed. Then it tells the display register to accept the "3" it's been staring at for so long. Since this register has a separate place for each of eight digits, it automatically puts the "3" in the far right-hand storage position. (Remember—both the "3" and the note in the flag register are in the special code that can be handled electronically.) The segment decoder immediately pounces on the code for "3," and begins energizing the necessary segments at the right times to show us a "3" in the display.

Why do simple things appear to get complicated?

Bingo! Finally, after dozens of instruction cycles, as the controller methodically obeyed instructions and decided which instruction to follow next, our "3" has been entered. Now the controller happily goes back to its idle routine. After making sure all keys are released, so it won't enter another "3," it begins watching for the next keystroke. All of this has taken only about a thousandth of a second.

You're beginning to see now just how many different things must be done in a digital system to accomplish a fairly simple task and that they can be done very rapidly. As we go on through the book, you'll find that this is really the secret of success for digital electronics. Every job and every number is broken down into very small steps and bits, so that it can be handled by very simple electronic circuits. We can put so many thousands of these simple circuits onto one integrated circuit chip that, working together, they can handle jobs and numbers as complicated as we need.

What happens on pressing the "5" and "plus" keys?

Let's move on rather quickly now through the rest of the addition problem, referring to *Figure 1-11* (which shows several more subsystems isolated from *Figure 1-6*).

Once the controller has confirmed that a "plus" command has been received, it checks the flag register to see if any earlier commands should be performed before it stores what's in the display in an operand register.

When we press "plus," the encoder tells the controller about it, and the controller in turn checks and verifies that a key was pressed as before. Recognizing that it has received an addition command rather than another digit signal, the controller then checks the flag register for any mathematical operations keyed in earlier that must be performed before the addition. Finding none, and because of the addition command, the controller makes the "routing subsystem" copy the "3" that's in the display register into the "operand register." In other words, it transfers the "3" to the operand register—which will now remember it. The operand register is identical to the display register and the accumulator register—which we will come to in a moment. All three registers are for storing an eight-digit number with decimal point and minus sign, if any.

Pressing the "5" key, in turn, triggers the same routine the "3" key caused. The display register is cleared and the "5" is stored, and a note is stored in the flag register to the effect that a new number has been entered. Now we have the copied "3" in the operand register and the new "5" in the display register *(Figure 1-11)*.

**Figure 1-11.
Entering "3 Plus 5" from
the Keyboard**

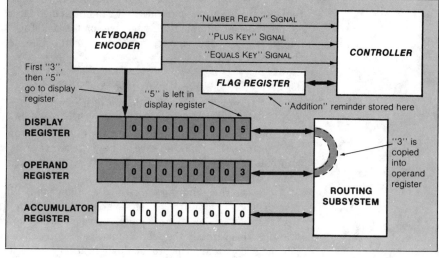

What happens when we press the "equals" key?

With both numbers of the problem electronically coded and stored in the appropriate registers, the controller coordinates the sequence of instructions that perform the add operation within the adder-subtracter subsystem.

Finally, when we press the "equals" key, the encoder tells the controller. The controller in turn checks and verifies the signal, recognizes that it has received an "equals" (or "end-of-problem") command, and checks the flag register to find out which operation it has to perform.

The addition note recovered from the flag register leads the controller to a next instruction that begins a programmed sequence of instructions—a "routine"—in this case, an "add" routine. (See *Figure 1-12*.) Instruction-cycle by instruction-cycle, the controller makes the routing subsystem transfer a copy of the electronically coded "5" from the display register, and simultaneously transfers a copy of the "3" from the operand register. Both numbers go to the "adder-subtracter" subsystem to be added.

**Figure 1-12.
Routing of Numbers
During Addition Process**

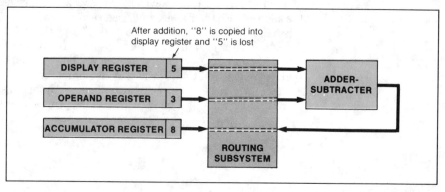

The adder-subtracter is a unit that handles all the arithmetic in the calculator. All it can do (and as we will see later, all it *needs* to do even to multiply and divide) is just what its name says—add and subtract.

The exact details of electronically adding numbers will be covered later in the book. But suffice it to say now that in one instruction cycle, the "5" and the "3" are added and the electronically coded sum of "8" is put into the accumulator register. Further instruction cycles figure out the proper decimal point and sign (plus or minus) for the sum, and then transfer it to the display register. There, the display sequence illuminates the "8" in the calculator display. And all these things triggered by the "equals" key happened within the IC seemingly faster than you could push the keys!

You didn't really know it, but the "5" that was in the display register was cleared out and lost, and by now the controller is back to the "idle routine," waiting for the next keystroke. At last then, the sum of "8" is in the display; and although the "5" is lost, the "3" is still remembered (stored) in the operand register in case we need it for further arithmetic operations.

How are decimal points handled?

We didn't mention it, but there was another routine that the controller had to do to make sure that we added our numbers correctly. It had to check the position of the decimal points in the numbers that were added and make sure the adder-subtracter had the decimal points "lined up" properly for addition.

We'll study this matter of handling decimal points further at a later time. But for now, as we see in *Figure 1-13*, let's just say that there is an electronically coded digit in a special position in each register where a number is stored, that tells where the decimal point is in the number. In *Figure 1-13*, for all numbers, the "0" for the decimal-point digit means the decimal points go at the far right of the stored numbers.

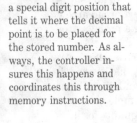

Each storage register has a special digit position that tells it where the decimal point is to be placed for the stored number. As always, the controller insures this happens and coordinates this through memory instructions.

**Figure 1-13.
A Separate Digit
Controls Decimal Point**

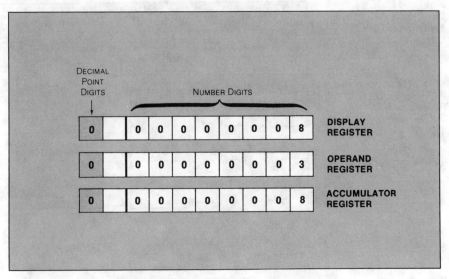

How can electricity transmit numbers?

So there you have a general picture of how the calculator goes about its business. You've seen all the major parts of the system, and how they work together. We've come a long way in understanding many of the things a typical digital system does.

Or course, we have not covered all the possible complications—such as entering decimal points and minus signs. And we haven't covered subtraction, multiplication, and division. But the fact is that all these matters are handled by the very same subsystems we have already watched in operation. They're managed by appropriate steps, done one at a time, in accordance with programmed sequences of instructions.

We'll understand more about these operations when we get further into the book. But for now, let's move on to finding out just *how numbers are represented* in a digital system such as this. We have seen where the "3" and the "5" go inside the calculator IC chip, and we mentioned that the numbers were coded so they could be handled electronically—but just what do they look like inside the IC?

To be specific, let's zero in on the connection between the *keyboard encoder and the display register*, back in *Figure 1-11.* We know the encoder generates an electronic code that represents numbers from zero up to nine, corresponding to number keys. In *Figure 1-11*, we see a broad arrow leading from the encoder to the display register, indicating a pathway for numbers. So what is this pathway like, and how does it work?

The answer – and the reason for it—goes back to our earlier discovery that the secret of success for digital electronics is that every job and every number is broken down into small, simple steps and bits. This is so that the tasks and information can be handled by very simple electronic circuits of the sort that can easily be put together in great quantities in integrated circuits.

**Figure 1-14.
Electrically Controlled
Switches as Switching
Circuits**

The ON-OFF switch function is the Basis of digital systems.

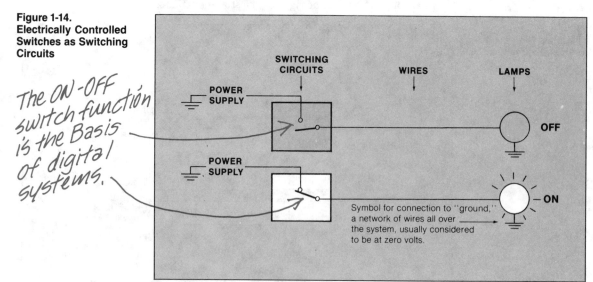

UNDERSTANDING DIGITAL ELECTRONICS

What is the simplest kind of electric circiut?

One of the most impor-
tant, fundamental con-
cepts of digital electronics
is the use of simple on-off
circuitry, producing binary
levels or states, which is
the basis of all digital
functions.

Now the simplest sort of electronic circuit is one that just
switches electricity on and off in a wire, as you turn a lamp on and off
with a switch. (See *Figure 1-14*.) We'll be studying such circuits in the
next chapter, and we will find that they use *transistors* rather than the
mechanical switches depicted in *Figure 1-14*. But for now, we'll just say
that switching circuits are never *part-way* on. The wires they control are
always clearly in one state or the other—on or off, high voltage or low,
large current or small, and so forth.

**Figure 1-15.
Switching Circuits Can
Receive or Transmit**

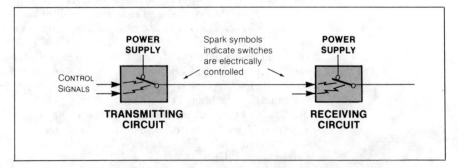

Now, as indicated in *Figure 1-15*, the kind of switching circuits
we're talking about are *controlled* by one or more input signals, which are
themselves either on or off. This means they can *receive* signals from *other*
switching circuits. And this is how numbers and information are sent
from place to place in our calculator, and *every* digital system—by
switching-circuits turning one another on and off.

What information can a switch send?

Now wait a minute, you may say. What kind of information can
you send by turning a switch on and off? How can anything this simple
handle the complicated kind of information involved in digital systems?

Well—it's true that a switch can't say much. But it can say
something. For a specific example, look at *Figure 1-16*. Two numbers, A
and B, are being "compared" by the adder-subtracter to see whether or
not A is greater than B (a job which the adder-subtracter handles simply
by subtraction). By switching one wire on or off, the adder-subtracter can
tell the controller the answer. "On" means, "*Yes*, A is greater." And "off"
means, "*No*, A is not greater."

**Figure 1-16.
Digital Information on
One Wire**

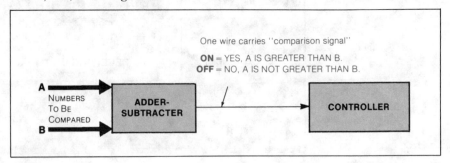

Each binary output represents a bit of information that has one of two values. By transmitting several bits of information at the same time, in a predetermined code, complex messages may be sent between the calculator's subsystems.

This is an example of the *basic unit of information* in all digital systems—the very *simplest* possible statement that can be made. It's just a specification of one out of two alternatives—a matter of yes or no. We call this amount of information one "bit." One reason this is a good name is that a bit represents the *smallest possible piece of information*. What we have discovered, then, is that at any one moment, a switch can transmit *one bit of information*.

But how can a switch sending one bit help us with the problem we're attacking about how to transmit *numbers?* What number can you send with *one bit* of information? We could let "off" represent "one," and let "on" represent "two"—but what good would that do ?

The answer is that to transmit larger numbers than one or two, we simply use *more than one wire*. This will give us a lot of different combinations or patterns of "on" and "off," and we can let each combination represent a different number, according to some sort of code.

Figure 1-17 shows specifically how numbers are transmitted in our example calculator. The "transmitting unit" above and the "receiving unit" below represent the encoder and the display register back in *Figure 1-11*. Furthermore, these same units represent *any* two subsystems that transmit and receive numbers. They all work the same way in this particular calculator.

Figure 1-17.
Using Multiple Wires to
Transmit Numbers

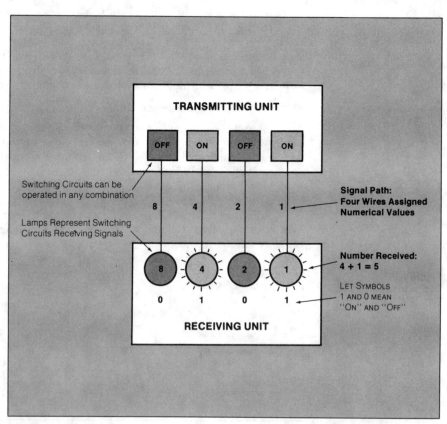

UNDERSTANDING DIGITAL ELECTRONICS

What's an example of a code used for numbers?

The code used in calculators is a combination of bits with either a one (1) or a zero (0) value in specified digit positions. The ones represent the circuit in the on or closed state, the zeros, the off or open state. Each one or zero is discrete and not read as you would a familiar decimal number.

As you can see in *Figure 1-17*, the example number code we're using consists of letting each wire represent a number: 8, 4, 2, and 1. The number transmitted is just the *sum* of the numbers represented by the wires that are switched on. At the particular moment illustrated in the figure, the "4" wire and the "1" wire are on, so the number being transmitted is *five*. (We're pretending the receiving unit has little lamps to show us which lines are "on.")

Below the lamps, you see a string of four symbols: 0101. These symbols, zero and one, provide a handy way that's used throughout the field of digital technology, to indicate whether a wire is on or off. We're going to let *zero mean "off,"* and let *one mean "on."* So in this particular code scheme, 0101 always means "five." We read it as "zero-one-zero-one." It's *not* "a hundred and one."

As for the rest of the combinations used to represent the numerals zero through nine, they're shown in *Figure 1-18*. You may recognize this code scheme as consisting of *binary numbers*. It's the most common code scheme used in digital systems. There are others, but we'll defer further discussion of code schemes to a more appropriate time.

**Figure 1-18.
A Calculator Binary
Number Code**

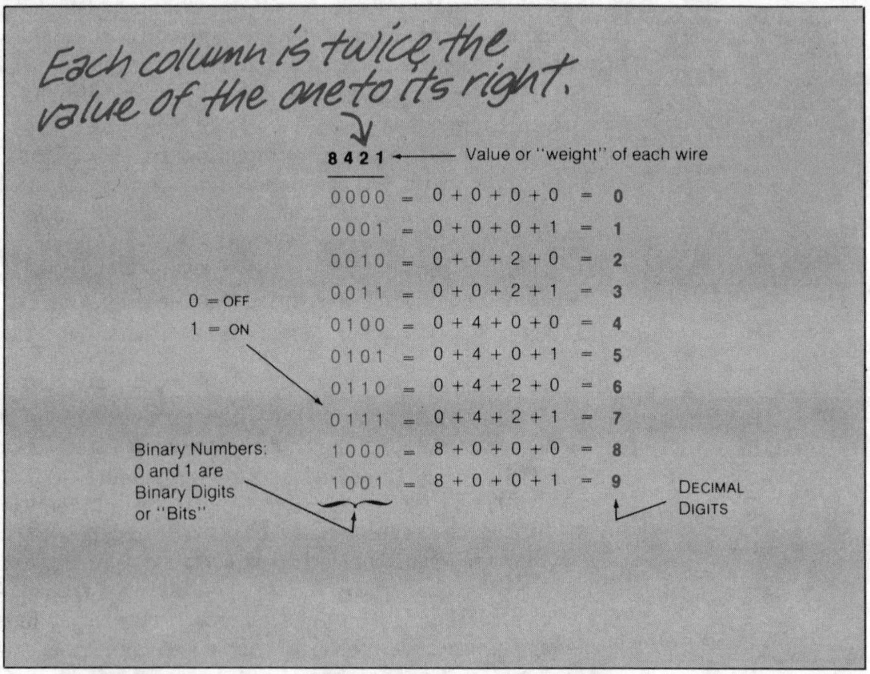

How do "binary numbers" show us what "digital" means?

This business about "binary numbers" in *Figure 1-18* will show us where we get the word "digital," as in "digital systems." Let's start by considering how our everyday "Arabic" number system works.

In writing an Arabic number, we use ten different symbols: 0, 1, 2, 3, 4, 5, 6, 7, 8, and 9. These symbols are called numerals, or *digits*. A digit is a position in a number telling how many ones, how many tens, how many hundreds, and so forth. This system is also called the "decimal" system, and the numerals are called the "decimal digits." Decimal means something related to the number ten.

The decimal number system uses ten values (0 to 9) to represent digits in a number; the binary system uses two (1 and 0). In digital circuits, the two binary values are represented by two different levels of electrical voltage or current, one level assigned to each value.

Now "binary" means something related to the number *two*, or something with two parts. In writing a number using the binary system, we only use *two* symbols, 0 and 1. These numerals are the *binary digits*. Each position for a digit in a binary number stands for *twice* what the next position to the right stands for. So that's what binary numbers are.

And by the way, the first and last letters of "binary digit" are where we get the word "bit." (Remember, a bit is the basic unit of information in all digital systems—the smallest possible piece of information.) A bit is a binary digit—a 0 or a 1.

And so a digital system is one that uses *digits* for all the information it handles. Even information that has nothing to do with numbers is reduced to the *form* of numbers using special codes, and the codes are made out of *digits*.

You can see that this definition does not limit digital systems to those that use *binary* numbers. For example, old-fashioned mechanical adding machines are digital systems that use *decimal* digits. They represent each position in a number by a gear or bar with ten teeth, so that it can be set at any one of ten different positions.

To represent numbers purely in decimal form by using electricity, you would need a different voltage level for each of the ten digits. Switching circuits that can handle ten different voltage levels are pretty expensive, however. So all modern digital electronic systems use *binary* digits (zeros and ones), as represented by very simple electronic circuits switching on and off. Consequently, whenever we say "digital system" nowadays, we take it for granted that we're speaking of *binary* digital systems.

What are the four principal functions in digital systems?

Let's press ahead now with one more detail of our initial understanding of digital electronics. Remember we said early in the chapter that to begin our understanding of a digital system by studying a calculator, we have to learn four things: How numbers are *transmitted* as inputs from the keyboard, how they are *stored*, how they are *added*, and how they are formed on the *display* as an output.

We're already covered the questions of transmission, addition, and display. (We haven't yet seen *exactly* how the adder-subtracter manages to add, but we've gotten a good overview of how the system goes about *making* the addition take place, and that's good enough for now.) So the remaining feature to cover now is *storage*.

If you look back at the entire calculator system shown in *Figure 1-6,* you'll recognize several different subsystems that we've talked about which *store* numbers and other information. The three number registers store numbers, the flag register stores miscellaneous notes, the address register stores numerical instruction addresses, and the instruction register stores instructions. Well, how can switching circuits store information?

**Figure 1-19.
A Switching Circuit That
Stores or Remembers
Information**

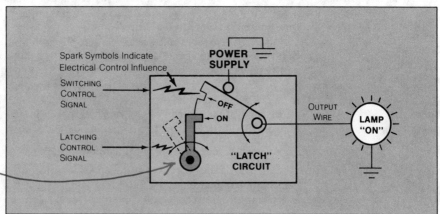

The latch will not change its status until it receives a latch control signal.

A latch is a common electronic storage circuit that stores (latches) the value of a bit on its input upon receipt of a latching control signal, and holds it until a new latch control signal stores new data.

How can a switching circuit store information?

We'll see in detail in later chapters how storage-type switching circuits work. For now, however, let's be content with a general mental picture of what's going on in a storage-type (or memory-type) switching circuit.

Figure 1-19 represents a type of circuit called a "latch." It has this name because the output can literally be *latched,* or fastened, in one state or the other—on or off. We've indicated this by showing an actual mechanical latch or hook, engaging either of two notches (labelled "on" and "off") in a pivoting mechanical switch element. Thus, the switch can be latched either in the "on" position, so that electricity flows from the power supply to the lamp—or in the "off" position.

This picture is patterned after the switching circuits we studied back in *Figure 1-15.* Actual electronic latches, of course, use transistors rather than mechanical parts—but the results are much the same.

Let's consider what this circuit will do with the latch lever *retracted,* the dotted line position in *Figure 1-19.* As before, we're using little spark symbols to indicate the control function. The switch element is turned off or on by the incoming "switching control signal." With the latch retracted, it will change each time the control signal changes. Now if the latch lever is *engaged,* the switch is restricted from changing and will remain in the latched position until the latch is retracted again. The latch lever is either engaged or retracted by the incoming "latching control" signal being switched on or off. This is being done by a switching unit feeding the "latch control signal" wire.

Several latches combined
and functioning as one unit
are called a register.

Thus, to *remember* whether the switching control signal was on or
off at a certain moment—to *store this one bit of information*—we simply
latch this circuit in that state at that moment. And the switch stays there,
no matter how the switching control signal may change afterward, until
we release the latch again. Then as soon as the switching control signal
changes the switch, the stored information is lost or "forgotten," as the
output now represents new information to be latched.

Now if we represent this circuit as a digital electronic circuit that
is receiving one of the wires from the transmitting unit of Figure 1-17,
then putting four of these circuits together provides a storage unit of four
bits that can represent numbers from zero through nine as we showed in
Figure 1-18. We need only add a latch control signal line to the receiving
unit to make this so.

Such a receiving unit with four stages then would be a
register—which, of course, we have talked about a great deal in our
calculator system description. It would store *four* bits and hold them
temporarily until we want to change them.

How does a calculator represent all digital systems?

That's the end of our initial overview of digital systems, in the
form of an example calculator. Before we move on to the next chapter,
let's pause and think about how the ideas we've seen can be *generalized*
from the example calculator to digital systems at large.

The most important generality, as indicated earlier, is that all
modern digital electronic systems operate like the calculator by reducing
information and tasks to very *simple terms*—to a matter of on or off, yes
or no, 1 or 0. To handle information and tasks of any complexity whatever
requires employing *large quantities* of such simple statements and tasks,
doing it rapidly using *code schemes* by which many simple pieces of
information can represent a more complex bundle of information. You'll
see this pattern in every digital electronic system.

How does electricity suit digital system requirements?

Now we haven't made a point of it yet, but the fact is that you
can build a perfectly functional digital system *without using electricity at
all*. Nothing in our definition of digital systems says anything about
electricity—just about breaking information into little pieces, about using
numerical digits, and so on. One example we've already cited of a
non-electrical system—even a non-*binary* one—is a mechanical calculator.
Another example, a more up-to-date one, is certain binary digital systems
employing devices that switch *liquids or gases* flowing in little tubes. We
call these "fluidic" systems.

But the reason that *electricity* has been employed for digital
systems so successfully is that *electrical switching circuits*—which are
relatively simple and inexpensive compared to some other electrical
circuits—can be used to handle the very simple information and tasks

involved in binary digital systems. These circuits are the fastest, most convenient method we know for such purposes.

Why do integrated circuits fit in so well?

Integrated circuits are ide-ally suited to operate in a digital format. They can combine thousands of microscopic digital circuit switches to perform the most complex functions in an incredibly tiny area at tremendous speeds.

The first digital electric systems used electromechanical relays that actually contained little mechanical switches of the sort we have been imagining in switching circuits. Later digital electric systems used vacuum tubes instead. Soon the transistor came along as a replacement, and then semiconductor integrated circuits.

And here again we seem to have a marriage made in heaven. As we will see more clearly later on in the book, *integrated circuits are naturally adapted* to reducing simple switching circuits to microscopically small size, and packing countless thousands of them into an unbelievably small space, lowering the cost per circuit significantly.

This capability throws ICs right into the arms of digital systems—which as we have seen involve *many* simple tasks and pieces of information. Integrated semiconductor electronics is the best way we have found yet to implement digital systems—and it's getting better all the time as integrated circuit technology improves so that more and more circuitry is put on one piece of silicon material.

What do all systems do?

From here, we can move on to one more, even grander generalization drawn from our calculator example, illustrated in *Figure 1-20*. This generalization is made up of two ideas. First, the only things that any system does, or can do, are to manipulate information and do work (or both). That is, all that's going on in any system is the handling of various forms of information, perhaps associated with the doing of work.

**Figure 1-20.
The Universal System
Organization**

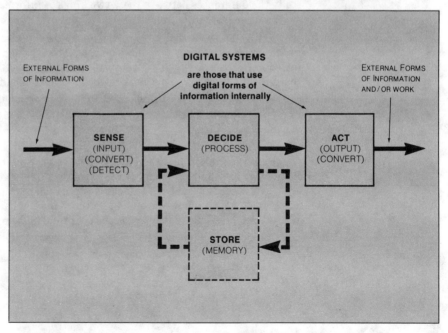

How are all systems organized?

All digital systems are organized and function in a similar manner. They sense incoming information and convert it for use by the system; make decisions on the input information and some other information that may be stored in the system; and output action dictated by the system.

And second, all systems are organized in the same fashion. They do their jobs in the same general steps or stages. First, they *sense* (or detect, or accept) information in various forms from the outside world, and convert it to forms of information that can be handled in the system. Then they make *decisions* based on this input information—meaning they process or manipulate the information. In doing so, they may *store* or remember some of the information for a time, or process it as a result of other information *stored* permanently. And finally, they take the resultant new information and *act* on the outside world with it—by converting it into external forms of information again, and perhaps by exerting some controlled form of work or energy. Think of any system you like, and this universal organizational concept can be construed to apply to it.

For example, our calculator's keyboard and encoder *sense* information and convert it into an internal form. Various subsystems *decide* and *store*. And the segment decoder and display system convert the resulting internal information into the desired *action* of showing you numbers in the display. This "digital electronic" system, of course, is handling the information in digital form.

How does this distinguish digital systems from others?

The significance of this universal system concept is that it shows us that *digital systems* are those that manipulate *information in digital form*, which we have seen means in the form of digits—little separate pieces of information. There's only one other general method for handling information, and it's called "analog." In Chapter 5, we'll study the differences between these two kinds of information, and the two kinds of system that result.

And now we really *have* come a long way! We've moved from a general understanding of a hand-held calculator, through an introduction to concepts of digital systems, to a grasp of the unifying concepts of all systems. This will provide a background of understanding as we proceed to dig into digital systems and see *how* they do the things we've been discussing.

Take a break

As you come to the end of each chapter, it will be a good idea for you to stop and take a breather. And before moving on to the next chapter, go back and study any of the parts that weren't clear to you at first. This is because a lot of the ideas covered in each chapter are necessary for your comprehension of material in later chapters. The glossary, and the quiz that follows each chapter, will help you review.

Quiz for Chapter 1

1. How does the calculator circuitry know which key caused a signal in a keyboard input line?
   a. There's a different input line for each key.
   b. There's a different scan line for each key.
   c. By noting which scan line is on when the signal is received.
   d. B and C above.

2. Why do the numerals in the display flicker (although faster than you can see)?
   a. They use alternating current.
   b. They're off while the controller re-checks the inputs to verify a signal was received.
   c. The segment outputs can transmit only one numeral at a time.
   d. The display register only stores one digit at a time.

3. How is the controller able to do so many different things at different times?
   a. It contains a special, different circuit for each job it has to control.
   b. It really doesn't control the other subsystems – they pretty much act independently and automatically.
   c. It just repeats the same process for each job it has to do.
   d. It's told what to do by instructions fetched from the microprogram memory.

4. How are operations in all subsystems kept in step together?
   a. Each subsystem has a little "clock" unit.
   b. By control signals from the controller.
   c. By signals in the scan lines.
   d. By timing pulses in three networks called phases.

5. When the "equals" key is pressed, how does the controller know which arithmetic operation to perform?
   a. It checks a note it made about this in the flag register.
   b. The current microprogram instruction contains this information.
   c. There's a place in each number register for minus signs, plus signs, multiplication signs, and so forth.
   d. It has already performed the necessary operation and is just waiting to display the result.

6. All the arithmetic in the calculator is handled by a unit that can only:
   a. Add
   b. Subtract
   c. Compare two numbers
   d. A and B above

7. The switching circuit
controlling a wire in a binary
digital system:
   a. Is either "on" or "off."
   b. Is often "part-way" on or
   off.
   c. Can be controlled by other
   Switching circuits.
   d. A and C above.

8. A binary switching circuit can
indicate a choice between how
many alternatives (in one
wire at one moment)?
   a. One
   b. Two
   c. Ten
   d. Depends on the circuit
   design.

9. What is a "bit" in a digital
system?
   a. A binary digit (1 or 0).
   b. The basic unit of
   information.
   c. The smallest possible piece
   of information.
   d. All of the above.

10. Where do we get the name
"digital" electronics?
   a. You key in numbers with
   your fingers (digits).
   b. All digital systems use
   binary digits (bits).
   c. All digital systems use
   some sort of numberical
   digits (decimal, binary,
   etc.)
   d. All digital systems have
   digital number displays as
   in the calculator.

11. Which binary number
represents "seven?"
   a. 1111111
   b. 7
   c. 0777
   d. 0111

12. Which of the following
manipulate information and
possibly do work?
   a. All systems.
   b. Only digital systems.
   c. Only binary digital
   systems.
   d. Only electronic binary
   digital systems.

13. Which do all systems have in
common?
   a. Sensing external
   information.
   b. Making decisions and
   possibly storing
   information.
   c. Acting to produce external
   information and possibly
   work.
   d. All of the above.

14. Digital systems are those
which:
   a. Sense, decide, store, and
   act.
   b. Manipulate information
   and do work.
   c. Handle information in
   digital form internally.
   d. Deal with digital
   information in the external
   world.

(Answers in back of the book)

How Digital Circuits
Make Decisions

As we begin a new stage in our learning process, let's remind ourselves of what we have covered already, and why we did so.

First, we gained a general familiarity with the operation of a simple hand-held calculator. Of all digital systems, the calculator is perhaps the most familiar and intriguing, so it provided a good way to get us into this subject.

And indeed we already are into the subject with both feet. Based on our study of the calculator, we have grasped the basic organizing principles that are common to all electronic digital systems. Some of these principles apply also to all digital systems, whether electronic or not. And some even encompass anything that we can call a "system" – even if it's not a digital system.

And this is where we're going to pick up the subject now – with the universal system organization that we learned in Chapter 1, as we applied it to the hand-held calculator. This will lead us into our topic for this chapter.

How does the universal organization apply to the calculator?

All of the parts and subsystems of a calculator can be placed in one of four functional categories: sense, decide, act, or store.

Figure 2-1 shows how the various parts and subsystems of the calculator are categorized according to which of the "universal functions" they mainly perform – based on whether their primary job is to sense, to decide, to store, or to act. (There's actually a certain amount of decision-making involved in *all four* stages – but decisions are the *main* job only in the "decide" section.)

**Figure 2-1.
Calculator Subsystems
in Universal System
Form**

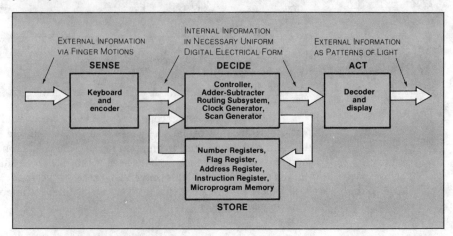

Chances are you pretty well understand how the switches in the keyboard *sense* external information from your fingertips, and how the light-emitting diode display *acts* to produce new external information in the form of patterns of light. And our initial picture of how switching circuits can *store* information is probably fairly satisfying to you for the time being. But you probably have some pretty big question marks in your head with regard to the *decide* function. How in the world can electrical circuits actually make *decisions?* Can it be that electric circuits have some form of intelligence?

Well, of course, they don't. But this is indeed a natural question. And it's such a crucial question for digital systems that we're going to devote this entire chapter to it.

What's the simplest example of decisions in the calculator?

The simplest decisions in the calculator are made in the keyboard encoder of the "sense" block of the calculator rather than in the "decide" stage. The "sense" stage not only detects incoming signals, but also must make simple decisions when converting these inputs to a form that's usable by other circuits.

Looking back at *Figure 2-1*, then, let's pick a *very simple* decision-making unit as an example to study, to help us grasp the main idea of how digital circuits make decisions. Surprisingly enough, the simplest example is *not* in the "decide" stage (We'll postpone studying these more complicated subsystems until later). Instead, the simplest decisions are made in the *keyboard encoder*, over in the "sense" stage. *Figure 2-2* reminds us of what the encoder's job is, and why this decision-making unit is classified in the "sense" stage rather than in the "decide" stage.

**Figure 2-2.
Sensing Input Signals
from Calculator Keys**

To convert keystrokes to useable signals requires this "sense" unit to make decisions.

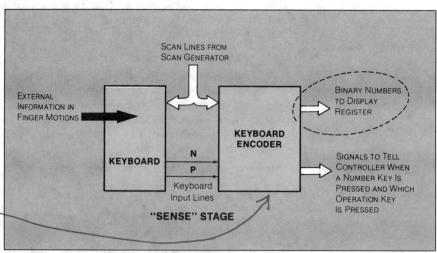

The "sense" stage, depicted in *Figure 2-2*, not only senses or detects external information by means of the switches in the keyboard — but it also *converts* this information into a form that's convenient for the other subsystems (which we have seen is the electronic "binary code") by means of the encoder. This conversion process involves *decisions*, as we will soon see — decisions that are very well suited for introducing us to how they are performed.

The process of converting key strokes to a number signal is performed in a two-step operation by the encoder. The first step determines which number key has been pressed; the second step determines the correct code wires to activate to send the desired number signal to the display register.

**Figure 2-3.
Converting Keyboard
Signals Into Numbers
with Encoder**

What steps are involved in encoding numbers?

First, let's narrow the scope of our study of the encoder. It will be sufficient for us to find out only how the encoder generates the *number signals* to the display register, shown as a broad arrow emphasized by a circle in *Figure 2-2.* We will take it for granted that the signals to the controller, shown further below, are produced in much the same fashion. (These signals tell the controller when a number key is pressed — without saying which one — and when each of the "operation" keys is pressed, such as plus, equals, and so forth.) So let's inquire into the decisions involved in encoding keystrokes into binary code.

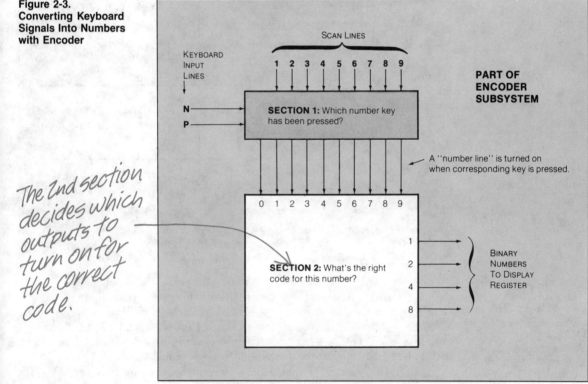

The 2nd section decides which outputs to turn on for the correct code.

The encoder generates numbers in two steps, and each step will illustrate a different kind of basic decision-making circuit for us. These steps are illustrated in *Figure 2-3* as two sections of the encoder.

In the first section, some circuits of one kind decide *which number key has been* pressed, according to which of the keyboard-input lines and scan lines are "on." The answer is transmitted by turning on one of ten "number lines" leading down to the second section. Down there, some circuits of the other kind decide *which of the four wires* leading to the display register to turn on, to transmit the number according to the binary code we learned back in *Figure 1-18.*

What does an AND gate do?

Let's look into the first section shown in *Figure 2-3*, and consider the switching circuit that decides when to turn on the "number one" wire leading down to the second section. *Figure 2-4* shows what this particular circuit has to do, and where it gets its input information. Let's consider this job carefully, because it's one of the most basic decisions in digital electronics.

**Figure 2-4.
Section 1 of Encoder**

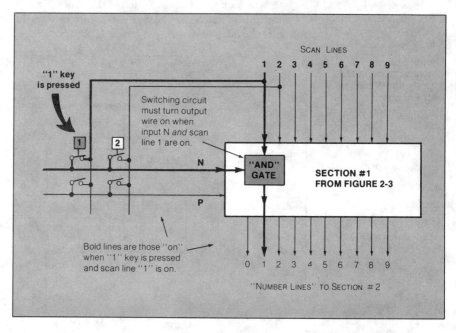

Electronic switching circuits called gates are the functional heart of digital electronics. A group of AND gates in the encoder decide which key has been pressed because the output of one AND gate will be turned on.

We want this circuit to turn on the "number one" wire whenever the "one" key is pressed on the keyboard. Remember now – pressing the "one" key causes keyboard input line N to be "on" when scan line 1 is "on." No other keys (such as the other three shown in *Figure 2-4*) will make *both* these inputs be on at the *same time*. Therefore, our "number one" switching circuit must turn on whenever both input N *and* scan line 1 are "on."

This circuit may be considered a "coincidence detector," because it responds only when it discovers both input signals "on" at the same time. (Two things happening at the same time are called "concident.") The circuit can also be considered to be like a *gate in a fence*, because an "on" signal in either input causes the output to be in the same state as the *other input*. This makes it seem as though a "gate has been opened up" for signals in the other input to "pass through." But an "off" signal in one input "shuts the gate" against signals in the other input, causing the output to remain "off." This idea is where we get the name for the circuit. It's called a "gate." And since there are other circuits also called gates (which we will see in a moment), this kind is called the AND gate, with the "and" spelled in capital letters.

How does an AND gate work?

Before we see how all the rest of the ten "number wires" are turned on by AND gates, let's look at the idea of how AND gates work, shown in *Figure 2-5*.

Regardless of the details of circuit design (which we will study in due time), all AND gates fit the mental picture presented in *Figure 2-5*. They all act as though they consisted of electrically-controlled switches connected "in series" as shown, with each switch turned on by an "on" signal in a particular input wire. ("In series" means with the same current passing through both switches.) In this particular example, when "on" signals in both input N and scan line 1 turn on both switches at the same time, electricity flows from the power supply to the output line. This gives us the output signals we want for each of the possible combinations of input signals, as summarized in the "function table" in the figure. It's as simple as that.

The AND gate will produce an "ON" output only when all input signals are "ON". All other input combinations produce an "OFF" output. The AND gate is one of the basic building blocks used to build any digital system.

Figure 2-5.
How an AND Gate Works

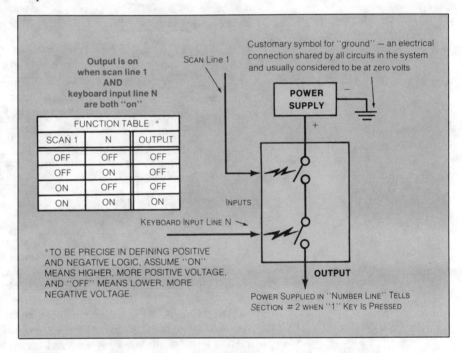

Output is on
when scan line 1
AND
keyboard input line N
are both "on"

SCAN LINE 1

Customary symbol for "ground" — an electrical connection shared by all circuits in the system and usually considered to be at zero volts

POWER SUPPLY

FUNCTION TABLE *		
SCAN 1	N	OUTPUT
OFF	OFF	OFF
OFF	ON	OFF
ON	OFF	OFF
ON	ON	ON

INPUTS

KEYBOARD INPUT LINE N

*TO BE PRECISE IN DEFINING POSITIVE AND NEGATIVE LOGIC, ASSUME "ON" MEANS HIGHER, MORE POSITIVE VOLTAGE, AND "OFF" MEANS LOWER, MORE NEGATIVE VOLTAGE.

OUTPUT

POWER SUPPLIED IN "NUMBER LINE" TELLS SECTION #2 WHEN "1" KEY IS PRESSED

Do you realize what you have just learned? Consider for a moment what a heavy idea has been revealed to you: ELECTRICALLY CONTROLLED SWITCHES CAN MAKE DECISIONS! We have wormed our way down to the very foundations of our calculator — and indeed, the foundations of *all* digital electronics. And down at the bottom, we have uncovered one of the building-blocks that all digital systems are made of. Connected together in the right patterns, large numbers of these AND gates — along with a few other very similar kinds of circuits — are what make every digital system work.

Again, we recognize the pattern we will see again and again in this field: *The decisions that a system is required to make can be broken up and subdivided into very simple decisions, which can be handled by very simple electric circuits* (or better still, electronic circuits).

What's the customary symbol for the AND gate?

Now in finishing up our explanation of how the first section of our number encoder works, we will need to show several AND gates in a small space. To avoid having to label each little box as an AND gate, we will use the customary symbol for AND gates, shown in *Figure 2-6*. Note that the output and all inputs are shown, but the power-supply connection is left off to keep the drawing simple.

To simplify illustrations, standard symbols have been assigned for gates — the one for the AND gate is shown. The truth table is a standard way to show a gate's output with all combinations of inputs. Notice that a gate may have more than two inputs.

**Figure 2-6.
Symbol and Definition of
AND Gate**

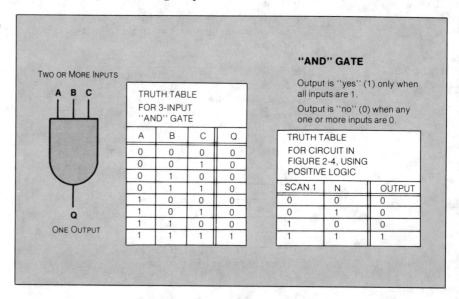

TWO OR MORE INPUTS

A B C

Q

ONE OUTPUT

TRUTH TABLE
FOR 3-INPUT
"AND" GATE

A	B	C	Q
0	0	0	0
0	0	1	0
0	1	0	0
0	1	1	0
1	0	0	0
1	0	1	0
1	1	0	0
1	1	1	1

"AND" GATE

Output is "yes" (1) only when all inputs are 1.

Output is "no" (0) when any one or more inputs are 0.

TRUTH TABLE
FOR CIRCUIT IN
FIGURE 2-4, USING
POSITIVE LOGIC

SCAN 1	N	OUTPUT
0	0	0
0	1	0
1	0	0
1	1	1

We're taking this opportunity to point out something new here in *Figure 2-6: three* inputs are shown on this gate (labelled A, B, and C), rather than two. This is to show you that an AND gate can have *more than two inputs*.

Another thing new in *Figure 2-6* is that we're showing the definition of an AND gate more *precisely* than before. An AND gate is actually defined in terms of the *logical meaning* of the inputs and outputs, in terms of the two basic bits of information a wire can carry, rather than in terms of the electricity itself. As we learned in Chapter 1, we call these bits 1 ("yes," or "true") and 0, ("no," or "false").

So to be precise, then, an AND gate is any circuit with two or more inputs and one output, whose output is 1 only when all the inputs are 1. The output is 0 when any one or more inputs are 0. The larger table in *Figure 2-6* shows what this means in the case of a 3-input AND gate. It's a list of all the possible input combinations and the resulting output for each combination. It's called the "truth table" for a 3-input AND gate.

How does our AND circuit qualify as an AND gate?

To determine the kind of gate and its truth table, the electrical conditions that represent the on and off conditions must be defined. Assigning the "ON" or 1 state the higher, more positive voltage and the "OFF" or 0 state , the lower, more negative voltage is called positive logic.

To prove that our AND circuit of *Figure 2-4* is *really* an AND gate according to the precise definition, we must first *decide what "on" and "off" mean in this particular application*. In the scan-line input, "on" means *yes*, electricity is being supplied only to the switches connected to scan line 1. Similarly, "on" in keyboard input N means *yes*, electricity is coming from one of the switches connected to this line. And electricity in the output means *yes*, the particular switch connected to both the input lines is turned on. The customary name for this situation (when "on" means 1 and "off" means 0) is "positive logic." (Actually, positive logic says that a *higher, more positive voltage* means 1 and that a *lower, more negative voltage* means 0. So to be accurate, let's *just* assume that our circuits are at a "higher" voltage when they're "on.") Anyway, using positive logic, we can write the truth table for the circuit of *Figure 2-4*. We just copy the function table from *Table 2-5*, writing 1 for "on" and 0 for "off." This truth table is shown at the far right in *Figure 2-6*. If we examine it, we find it's the truth table for a 2-input AND gate. So *in this application*, where we're using *positive logic*, our circuit qualifies officially as an AND gate.

What would this AND circuit be using negative logic?

If negative logic is used, the definitions are reversed, so the "ON" state is a 0 and the "OFF" state is a 1. This also reverses the truth table so that a 0 output will exist only if all inputs are 0. The type of logic must be known when analyzing a circuit.

On the other hand, suppose the circuit in *Figure 2-5* were used in some other application, where "on" ("high" voltage) meant "no" (0) and where "off" ("low" voltage) meant "yes" (1). The customary name for that situation is "negative logic." Then the truth table would be different. The output would be 0 only when both inputs are 0. So the circuit in that application *would not be an AND gate*. If you wanted a circuit to perform the AND function using negative logic, you'd have to build it another way.

And so, to be precise in our name for the circuit in *Figure 2-5*, we have to call it a *positive* AND gate. A negative AND gate would be different.

What does all this have to do with our explanation of the customary symbol for the AND gate in *Figure 2-6?* It means that whenever you use it to refer to a real electrical circuit, you have to say whether you're using positive or negative logic. That way, you say whether the symbol represents a positive AND gate or a negative AND gate. IN THIS BOOK, ALL SYMBOLIC DIAGRAMS USE POSITIVE LOGIC, unless we specifically say we're using negative logic.

How is the entire "first section" of the encoder designed?

Now that we have a convenient symbol for an AND gate, let's look at a diagram of the entire first section of the encoder, shown in *Figure 2-7* (using positive logic as before). There's a separate two-input AND gate for each "number line" running down to section two of the encoder. The inputs to each gate are simply the keyboard input line and the scan line that the particular key-switch is connected to, as shown back in *Figure 1-2*.

What you see in front of you in *Figure 2-7* is actually a miniature digital system. And you actually have learned how it works! From here on, the design of *any* digital system is just a matter, essentially, of hooking up the right kind of gates in the right way.

**Figure 2-7.
AND Gates Used to
Detect Number Key**

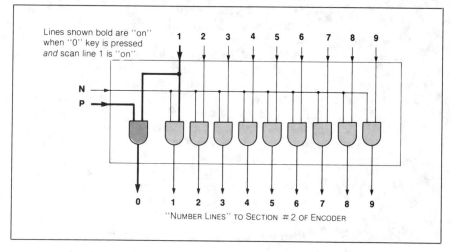

How can OR gates finish the decoder's job?

And speaking of the right kind of gates, there's another kind – just as important as the AND gate – that will let us design the second section of the encoder. You'll remember from *Figure 2-3* that the second section decides which of the four output wires to turn on to make a number for the display register. Well, this decision is made by several "OR gates." The entire design for encoding numbers – with both section 1 and section 2 lumped together – is shown in *Figure 2-8*.

The decision as to which display register wires to turn on to make a number is made by a group of OR gates.

**Figure 2-8.
Keyboard Circuit for
Encoding Numbers**

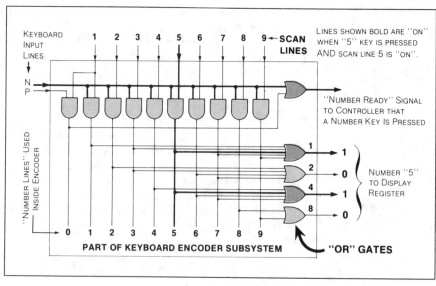

UNDERSTANDING DIGITAL ELECTRONICS

In the encoder, each OR gate receiving an "ON" signal from an AND gate produces an "ON" output. The four output lines together produce the binary code for the number to be displayed.

You can recognize the ten AND gates at the top of *Figure 2-8*, just as they appeared in the preceding figure. The additional OR gates are shown along the right edge of the subsystem. This spearhead shape is the customary symbol for an OR gate. (Since we're assuming positive logic as usual, the symbol means a *positive* OR gate.)

As to just what an OR gate does — you can get a clue by following the bold lines in *Figure 2-8*. These are the lines that are "on" (1) when the "five" key is pressed and 0101 is being encoded, which is "five" in the binary code listed back in *Figure 1-18*. (Remember—0101 means 4 plus 1 in the electronic code that we have decided to use to transmit numbers electrically.)

You already understand how one of the AND gates turns on the vertical "number 5" line inside the encoder. Now notice that this line is connected to two OR gates. It provides one input for the OR gate that turns on the "1" output line, and also one input for the OR gate that turns on the "4" output line. These are the gates we want to be "on" (1) to transmit a binary 5, right? Notice that when *any* digit key is pressed, a certain AND gate transmits a 1 to the OR gates that need to be turned on to make up the code for that number (4 and 1 for 5 . . . 4 and 2 and 1 for 7, and so forth). So can you see now what an OR gate does?

**Figure 2-9.
Symbol and Definition of
OR Gate**

The OR gate symbol has a spearhead shape. As shown in the truth table, any input of a 1 will cause the OR gate's output to be a 1. The OR gate output is 0 only when all of its inputs are 0.

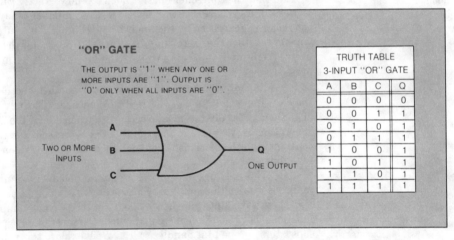

"OR" GATE

THE OUTPUT IS "1" WHEN ANY ONE OR MORE INPUTS ARE "1". OUTPUT IS "0" ONLY WHEN ALL INPUTS ARE "0".

TWO OR MORE INPUTS

A
B
C

Q

ONE OUTPUT

TRUTH TABLE 3-INPUT "OR" GATE			
A	B	C	Q
0	0	0	0
0	0	1	1
0	1	0	1
0	1	1	1
1	0	0	1
1	0	1	1
1	1	0	1
1	1	1	1

What does an OR gate do?

As shown in *Figure 2-9*, an OR gate's job is to transmit a 1 when *any one or more* of its inputs are 1. The output is 0 only when *all* inputs are 0. This action is summarized in the truth table shown in *Figure 2-9* for a 3-input OR gate.

This decision has the name OR because a 1 in this input OR that input OR the other input will give a 1 at the output. And it's called a "gate" because when you have one with only two inputs, and you hold one input at 0, the output is the same as the other input. Thus, you can "open" an OR gate (like a gate in a fence) by holding one input at 0, as you "open" an AND gate by holding one input at 1.

To follow an OR gate's action more closely, refer back to *Figure 2-8*, and look at the OR gate that turns on the "8" output. This OR gate's output is 1 when either the "number 8" line OR the "number 9" line is 1. And of course the reason we use these two number lines as inputs here is that eight and nine are the only numbers in our binary electronic code that have 1 in the "eights" place.

So that's what an OR gate does, and that's how number keys are encoded. However, if you've been studying *Figure 2-8* closely, there's a feature that may be puzzling you. The "zero" number line does not connect to *any* of the four OR gates that generate binary numbers, but instead ties into another OR gate higher up. Why is this?

What do we want the "zero" key to do?

The zero number line does not turn on any of the lines to the display register since the binary code 0000 is needed for the number. The "number ready" signal is sent to the controller when any number key is pressed.

Well — we *don't want* the "zero" key to cause a 1 in any of the binary output lines. To transmit a "zero", these four outputs must be 0000, right?

But what we *do* need the "zero" number line for is to participate in *alerting the controller* whenever one of the ten number keys is pressed, so the controller can go through the appropriate routines (discussed in Chapter 1) and tell the display register to store the number. Remember, we said that the controller doesn't need to know exactly *which* number key is pressed. As you can see in *Figure 2-8*, this "number ready" signal to the controller is generated by the uppermost OR gate. This output is 1 whenever either the "zero" number line is 1 (indicating the "zero" key is pressed) OR input line N is 1 (indicating one of the other number keys is pressed).

So there you have the design of our number-encoder network — the most important part of the keyboard encoder. The only other function of the complete subsystem is to recognize when one of the nine "operation" keys (plus, minus, equals, etc.) is pressed, and to tell the controller which one. This is done in much the same fashion as we have seen for numbers.

This network in *Figure 2-8* may not look very significant by comparison with the complexity of the entire calculator, or some other digital systems. But it illustrates not only the two most important gate functions, but also the principle of one of the most basic types of subsystems — one that handles *codes*. (We'll explore this principle in more depth in Chapter 3, with respect to the segment decoder subsystem.) You'll learn then that this general pattern of gates is repeated again and again in *many* different types of subsystems: a row of AND gates followed by a row of OR gates. Take a good look at it, because it's going to be with you for a long time.

How does an OR gate work?

As for the gates that the encoder has taught you about — you've already got a good mental picture of how to think of a circuit that acts as an AND gate in terms of positive logic ("high" voltage = 1, "low" voltage = 0). *Figure 2-5* showed us that all you need, essentially, is an electrically-controlled switch for each input (whether two, three, or more), with the switches arranged in series.

Figure 2-10 provides a similar mental picture of a circuit that's an OR gate when it's used with positive logic. Here again, we have an electrically-controlled switch for each input (as many inputs as you need, though only two are shown, labeled A and B.) But the switches are connected "in parallel" as we see here, rather than in series as in *Figure 2-5*. So the output would be "on" either when input A is "on" OR when input B is "on" (or when both are on, of course). You may imagine a lamp on the output to indicate when the output is "on."

The OR gate functions in a manner similar to electrical switches that are connected in parallel.

**Figure 2-10.
Operation of a Positive
OR Gate**

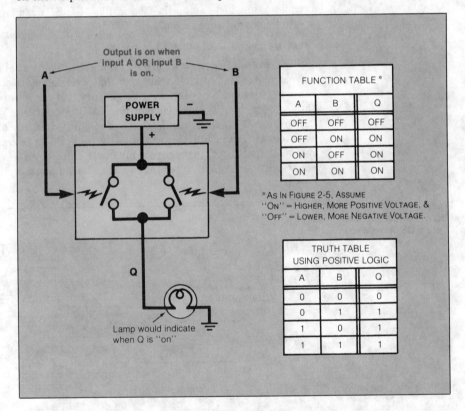

Output is on when
input A OR input B
is on.

A   B

POWER
SUPPLY

FUNCTION TABLE *		
A	B	Q
OFF	OFF	OFF
OFF	ON	ON
ON	OFF	ON
ON	ON	ON

* As In Figure 2-5, Assume
"On" = Higher, More Positive Voltage, &
"Off" = Lower, More Negative Voltage.

Q

Lamp would indicate
when Q is "on"

TRUTH TABLE USING POSITIVE LOGIC		
A	B	Q
0	0	0
0	1	1
1	0	1
1	1	1

The function table for this circuit is shown at top right. (The function table for *any* digital circuit shows the *electrical state* of the output for every possible combination of input states.) If we let "on" mean 1 and let "off" mean 0, the *function* table turns into the *truth table* on the right. As you can see, this is the truth table for a 2-input OR gate.

Do real gates work like this?

A real gate does not operate exactly like an electrical switch although the function is similar. Gates are made with transistors. The transistor whose operation is most like an electrical switch is the MOS transistor.

Now that you understand what AND gates and OR gates do and how they can team up to handle complicated decisions, let's move on to the subject of how *real* gates work – how transistors are hooked together to act as gates. As it turns out, transistors can't simply be inserted into the place of the switches in *Figures 2-5* and *2-10*, so let's see just how they *are* used. In doing so, we will be chipping into the bedrock *under* the foundation of digital electronics that we spoke of earlier. This bedrock is the semiconductor material of which transistors are made – which in these days is the silicon integrated-circuit chip.

Later on, we will see that gate circuits can use any of a number of different kinds of transistors. But we're going to limit ourselves right now to just *one* kind of transistor – the one that acts most nearly like the electrically-controlled switches we have been envisioning. (It's also the type used in most calculators.) This is the MOS transistor. "MOS" is pronounced by just saying the letters, like M-O-S. In a moment, we'll see where this name comes from.

What are the parts of an MOS transistor?

Figure 2-11 shows the general idea of the internal construction of the particular kind of MOS transistor that we're going to study. (The name of this kind is not important right now, but for your information, this is an "n-channel enhancement-mode MOS transistor.") What you see in *Figure 2-11* is a highly simplified and magnified picture of an area in an integrated circuit no bigger than a flyspeck. Down below is a pictorial reminder that this transistor roughly fits the idea of an electrically-controlled switch that we have been imagining in our gate circuits.

**Figure 2-11.
Internal Structure of an
"Off" MOS Transistor**

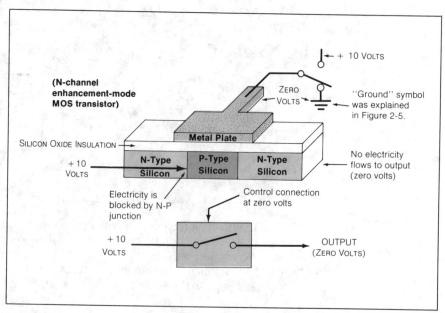

We won't go into semiconductor theory right now. All we'll say is that the main part of the transistor is a bar of silicon consisting of two different types: n-type and p-type. (Slight traces of certain other materials were added to the original pure silicon to make it either n-type or p-type — each type reacting differently to electricity.) Both ends of the bar are n-type silicon, and in the middle it's p-type.

On top of the silicon bar, there's a layer of silicon oxide. This is a substance much like glass. It acts as electrical insulation, so no electricity can pass through this layer. And above the oxide layer is a metal plate.

Now the two n-type areas are the main electrical terminals of our "switch". Electricity at ten volts is being supplied to the left terminal, trying to get through to the output terminal on the right. But at the left edge of the p-type region, the electricity is *blocked* by the most basic law of semiconductor action: electricity (positive electric charge, to be more specific) *cannot flow* across the border, or "junction", from n-type material to p-type. (This has to do with the n-type material having a more positive voltage in it than the p-type.) So the transistor is now in the "off" state.

The MOS transistor is based on a slice of silicon with P type material sandwiched between N type material. Across this sandwich is a metal plate which functions as the control terminal of the switch. The N material acts as the terminals of the switch. Current cannot flow between the N "terminals" when the control terminal has zero volts applied.

**Figure 2-12.
MOS Transistor in "On" State**

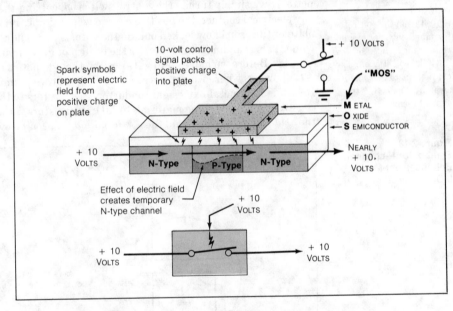

Now the metal plate acts as the control terminal of the "switch". As the schematic diagram indicates above in *Figure 2-11*, the plate is being held at zero volts by a two-way switch. Let's see how this switch applies a control signal to turn the transistor on.

How do you turn an MOS transistor on?

Figure 2-12 shows what happens when we flip the controlling switch up above to ten volts. The voltage pressure packs positive electric charge into the metal plate. Here, the charge finds itself at a "dead end" because it cannot pass through the oxide insulation.

When a positive voltage is applied to the control terminal, an electric field is created in the P material. Part of the P material is changed to form an electrical channel with properties like the N material on either side. Current flows from one N terminal through the channel to the other N terminal, thus, the "switch" is turned on.

However, the positive charge creates an *electric field*, which *does* exert itself right through the oxide, as a magnet exerts a field through a sheet of paper to a nail on the other side. This field is suggested by the little spark symbols. (There are no actual sparks involved – just an electric field.)

The effect of this electric field on the p-type region is one of the many remarkable things about semiconductors. As if by magic, the upper part of the p-type gap seems to *turn into n-type* as long as the positive charge is maintained on the plate above. Thus, a *temporary n-type channel* is formed between the two n-type terminals, allowing current to flow, and in effect turning on the transistor. This fact is suggested by the "switch" drawing below.

How do you turn an MOS transistor off?

Now to turn the transistor off again, it's important to note that we have to flip the control switch *back to zero volts again*. This is because we have to provide a *path for positive charge* to drain out of the metal plate. If we simply turned the control switch *off*, cutting the plate off from any electrical contact, the positive charge would simply remain on the plate until it somehow leaked out. So the transistor wouldn't turn off for a while. This is an important fact for the design of gate circuits.

To turn off the transistor, the charge on the control terminal must be completely removed. If the control voltage is just removed from the control terminal, the charge will remain for awhile and the "switch" will stay on. The charge must be drained away by connecting the control terminal to zero volts (ground).

Before we move on to gate design, though, notice that *Figure 2-12* shows us where we get the name "MOS" that we promised to explain. It stands for "metal-oxide-semiconductor". This refers to the "sandwich" construction of the three materials of MOS transistors – the *metal* plate, the *oxide* layer between, and the *semiconductor* silicon below.

**Figure 2-13.
Symbol for N-channel
Enhancement-Mode
MOS Transistor**

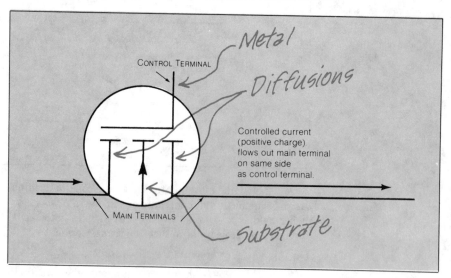

One thing we'll need in showing gate circuits is a *symbol* for our transistor. *Figure 2-13* shows the customary schematic symbol for the particular kind of MOS transistor we're using.

How does the transistor fit into a switching circuit?

To illustrate the idea of using our MOS transistor in a digital gate, we're first going to take up another switching circuit called the "inverter". (It's sometimes called the "NOT gate" – although it can't *gate* anything like the AND and OR gates can.) This is a building-block that's just as basic as the AND gate and the OR gate, although we didn't have an opportunity to show it in action in the encoder example.

**Figure 2-14.
Two MOS Inverter
Circuits**

The inverter is another of the building blocks and is used in combination with AND and OR gates to form logic circuits. Its function is to invert its input to the opposite state at its output; thus, a 1 input produces a 0 output and vice versa. The inverter is sometimes called a NOT gate.

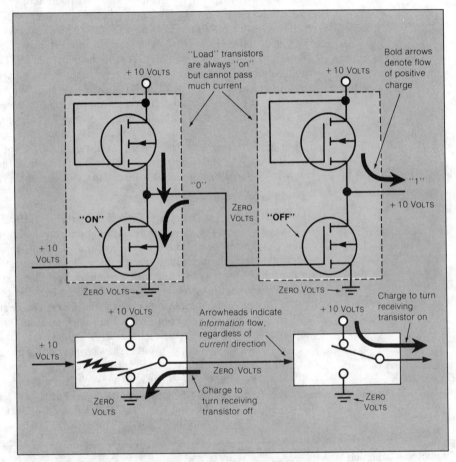

Figure 2-14 shows two MOS inverters in the two upper boxes, with the output of the first (left) providing the input for the second. The two boxes below give you a mental picture of what these inverters do, in terms of electrically-controlled switches. When an inverter input is at ten volts as shown on the left below, the output is connected to a "ground" terminal at zero volts. (The meaning of "ground" was explained back in *Figure 2-5*, although we didn't mention it in the text.) And when an inverter input is at zero volts as shown on the right, the output is connected to the "power supply" terminal above at ten volts.

These mental pictures below show us two things about the inverter. First, its purpose is to change, or "invert", the incoming signal to the opposite state. In our MOS circuitry, "on" (1) is ten volts or thereabouts, and "off" (0) is zero volts or something close to it. The inverter's output is always the "inverse" (the *opposite*) of the input. In simpler terms, we might say that the output is "flipped over" from the input.

The second point illustrated in the lower part of *Figure 2-14* is that an MOS switching circuit must not only be able to *supply* electricity through its output for the "on" state, but it must also be able to *drain electricity back into the output* for the "off" state. This is so the output can *turn off* an MOS transistor in the receiving circuit, as we have already learned in studying how an MOS transistor is turned off. This requirement of "two-way" output current applies not only to the inverter circuits in *Figure 2-14*, but to *all* MOS gates, which as we will see are constructed very much alike. In fact, it applies to *all electronic digital circuits*. Current must be able to move both ways in all signal wires.

By being able to drain current back through its output, the MOS inverter can drain the charge from the control terminal of the receiving transistor to turn it completely off.

Incidentally, even though electricity sometimes flows *into* an *output* of digital circuits, the thing that always flows *out* of an output is *information*. The small arrowheads on the signal lines in the diagram below in *Figure 2-14* represent information, not current.

How does an "inverter" circuit work?

Referring now to the schematic inverter circuits above in *Figure 2-14*, we can see how a 10-volt input signal turns on the lower transistor, connecting the output to the zero-volt ground as we desire. As for the *upper* transistor in each inverter (called the "load" transistor) — it is a very *specially-made* transistor with a comparatively long and narrow channel between the two n-type terminals. It is always kept "on" by having its control terminal connected to the 10-volt power supply. But it can't supply very *much* current — it chokes back most of the current like the electrical devices we call "resistors".

The inverter circuit is formed by two transistors connected in series. The upper transistor (the load transistor) acts like a resistor. It is always in the on condition and supplies voltage to turn on the receiving transistor when the bottom transistor is off. When the bottom transistor (the input transistor) is on, current is supplied through the load transistor from the power supply. The bottom transistor drains the current from the receiving transistor.

When the lower or "input" transistor is "off", the load transistor can supply *enough* current through the output to charge up the control plate of a receiving transistor in another circuit rapidly enough (very little current is required). But when the input transistor is "on", the small current supplied by the load transistor is not enough to interfere with the drainage of charge from the receiving transistor.

**Figure 2-15.
Symbol and Definition
for the NOT Gate**

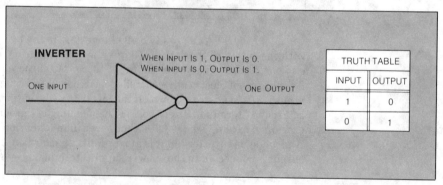

INVERTER

When Input Is 1, Output Is 0.
When Input Is 0, Output Is 1.

One Input   One Output

TRUTH TABLE	
INPUT	OUTPUT
1	0
0	1

Figure 2-15 shows the customary symbol used in diagrams for the inverter or NOT gate. Regardless of the internal circuitry, the output is always in the inverse (opposite) state from the input, as summarized by the truth table in *Figure 2-15*. (Obviously, this is so whether we're using positive or negative logic with a particular circuit.) Presently, we will be needing this symbol in a diagram of an MOS version of our keyboard encoder.

Another type of gate circuit, and one that is easier to build than AND and OR gates, is the NAND gate. In the NAND gate, the output is always in the high state unless all inputs are high. The effective function of the NAND gate is a NOT gate (inverter) in series with an AND gate.

What other gates can we build with MOS transistors?

Of course, our keyboard encoder (like the initial designs of most digital subsystems) needs AND gates and OR gates. However, it turns out that these gates are not so convenient to build out of MOS transistors as two *other* kinds of gate circuits. These other two are just as useful – in fact, *more* useful in many applications.

**Figure 2-16.
MOS NAND Gate and
Symbols**

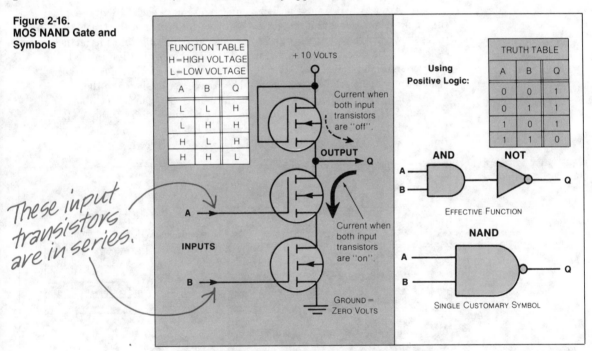

These input transistors are in series.

Figure 2-16 shows one of these other gate circuits. The schematic diagram at the left shows that this circuit consists essentially of an inverter (as in *Figure 2-14*) with not one but *two input transistors in series*. The bold arrow shows how the output is connected to ground only when both input transistors are "on". (Obviously, we could include more than two input transistors here if we need more than two inputs.) So the output can only be in the low-voltage state when all the input transistors are "on" – that is, when all the inputs are in the high-voltage state. Otherwise, if any one or more inputs are in the low-voltage state, the output is high-voltage.

Why do we use "H" and "L" in function tables?

The symbols H (high) and L (low) are more correct than on or off when constructing function tables for electronic digital circuits.

The action of this gate is shown in the function table left of the circuit in *Figure 2-16*, using "H" for the high voltage or more positive state (ten volts, we said earlier) and "L" for the low-voltage, more negative state (zero volts, or ground). These symbols are typical of function tables for all "real" electronic digital circuits (not just MOS circuits). We can't use "on" and "off" because the outputs are *always* on, in a sense — being connected either to the ground or to the power supply. What distinguishes one electrical state from the other is really the *voltage* in a wire. So "high" (H) and "low" (L) mean the same thing in *real* digital circuits as "on" and "off" did in the imaginary "electrically-controlled-switch" circuits that we have found so useful as a learning device.

What is a "NAND" gate?

Anyway — if we use positive logic and replace each "H" in the function table with a 1 and replace each "L" with a 0, we get the truth table shown at the right in *Figure 2-16*. This particular digital function is called "NAND" because the output is just the *inverse* of that for an AND gate. NAND stands for "NOT-AND". This truth table can be represented symbolically by an AND gate followed by an inverter as shown in the figure. The single customary symbol is a combination of the two symbols — it's an AND symbol with a little circle at the output to signify the AND function is inverted. Since the "real" circuit at the left performs the NAND function when using positive logic, it's a *positive* NAND gate.

**Figure 2-17.
MOS NOR Gate and
Symbols**

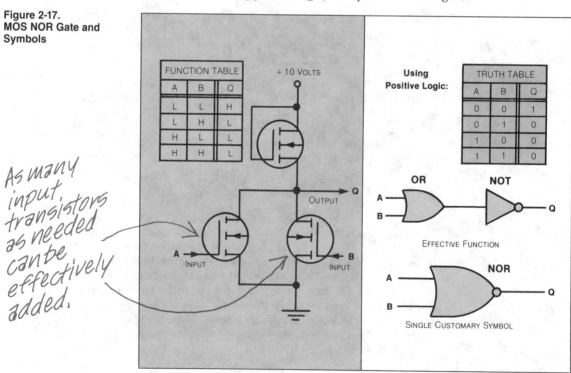

As many input transistors as needed can be effectively added.

What is a "NOR" gate?

The NOR gate will have a 1 output only when all inputs are 0. It functions like an OR gate in series with a NOT gate.

Figure 2-17 shows the other MOS gate that we referred to earlier, which is a positive NOR gate. The NOR function is described in the truth table on the right – the output is 1 only when all inputs are 0. Thus, the output is the inverse of the OR function. So a NOR gate acts as an OR gate followed by an inverter as shown below the truth table. The name "NOR" stands for "NOT-OR". As you can see, the customary symbol for a NOR gate is an OR symbol with an "inversion" circle at the output point.

It's easy to see how the circuit on the left acts as a *positive* NOR gate. Applying a "high" (1) signal to *either* input transistor will make it connect the output to ground. This produces a "low" (0) output signal. But if *neither* input transistor is on, the small continuous current through the load transistor provides a "high" (1) output signal. As with positive NAND gates, we can build positive NOR gates with as many input transistors as we want.

How can we use MOS gates in the encoder?

Finally now, we're ready to see how our keyboard encoder could be built using real MOS gates. *Figure 2-18* shows one way to do it (assuming positive logic, as usual).

What we've done here is to replace *all* the gates in our original design *(Figure 2-8)* with positive NAND gates. We've also added an inverter, which you can see. But the arrangement and connections of the wires are the same as before. Believe it or not, this new design provides correct signals in the five output lines!

You can prove this to yourself – at least in the case of the "5" key being pressed – by following all the bold lines, which are "on" (1). Remember – a NAND gate produces a 0 only when all inputs are 1. Otherwise, it produces a 1. Notice that the only number line that's 0 is number 5. This is because this is the only NAND gate with 1 at both inputs. The only two NAND gates that take an input signal from the number 5 line are those for the "1" output wire and the "4" output wire. Therefore these two are the only gates in the group of four that are receiving a 0 signal. Since the inputs to these two gates are not all "ones" like the other two gates, their outputs are 1.

Look up at the NAND gate producing the "number ready" signal. The inverter turns the 1 in input N into a 0 going into the NAND gate. Since both inputs to this gate are not 1, the output is 1.

This, then, is one way to build the encoder with the kind of gates that are available in the MOS family of integrated circuits, using positive NAND gates and an inverter. However, this circuit looks pretty confusing, because we can't follow the action as easily as we could with AND gates and OR gates. So let's see how to make better sense out of it.

The AND and OR gates of *Figure 2-8* can be replaced with NAND gates which are used in the real circuits. Read the text carefully for an explanation of how it works.

**Figure 2-18.
A Decoder with NAND
and NOT Gates**

How do you make sense out of a "NAND" gate arrangement?

Let's narrow our scope down to making sense out of the three NAND gates that act together to produce the signal in the "8" output wire in *Figure 2-18*. That's the very bottom NAND gate and the two above that receive scan lines number 8 and 9. These three NAND gates are shown by themselves in *Figure 2-19*.

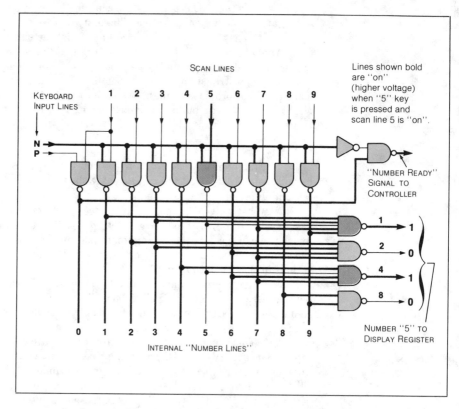

As shown in *Figure 2-19*, the first step in understanding what's really going on here is to *imagine* that each positive NAND gate is a positive AND gate and an inverter. Then imagine a box around the lower imaginary AND gate and all three inverters. Notice that inside this box, we've got *negative logic!* This is because a *low* voltage in a number line means *yes*, this key is pressed. And a low voltage in the AND gate output means yes, key 8 OR key 9 is pressed.

Now consider the positive AND gate in the box, and ask what it does in the case of negative logic. We know both its inputs must be high to make the output high. In terms of negative logic, that means both inputs must be 0 to make the output 0, since H = 0 and L = 1. If any input is 1, the output is 1. That should sound familiar to you. *It's the function of an OR gate.* We have discovered that A POSITIVE "AND" GATE IS ALSO A NEGATIVE "OR" GATE!

**Figure 2-19.
NAND Feeding NAND are
Equivalent to AND
Feeding OR**

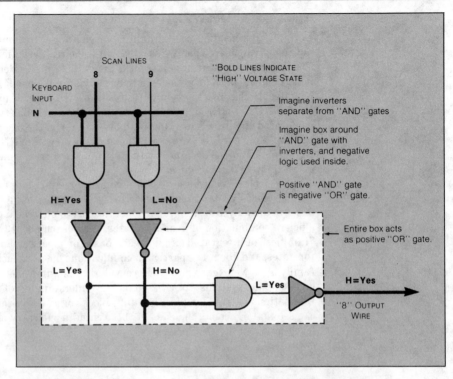

A positive AND gate is a
negative OR gate and vice
versa. A positive NAND
gate is a negative NOR
gate and vice versa.

In other words, the two electrically-controlled switches in series
back in *Figure 2-5* will be an OR gate if you write 1 for "off" and 0 for
"on" in the function table. You can also prove to yourself that a positive
OR gate (with switches in *parallel)* is a negative AND gate. Similarly, a
positive NAND gate is a negative NOR gate, and vice versa.

But let's get back to the imaginary box in *Figure 2-19. In our
mind,* we can replace this entire box — AND gate, inverters and all — with
a positive OR gate. That is, a positive AND gate with inverters on all
inputs and the output, when taken all together, make a positive OR gate.

Having gone through these mental conversions, you can see what
two positive NAND gates feeding a third really do. They act as two
positive AND gates feeding a positive OR gate. Or putting the same fact
another way — a positive NAND gate with inverted inputs acts as an OR
gate.

(You can also prove to yourself that three *negative* NAND gates
arranged as in *Figure 2-19* act as two negative AND gates followed by a
negative OR gate. And this means that the same thing holds for both kinds
of NOR gates. They act as OR gates followed by AND gates.)

Looking back at *Figure 2-18*, you should understand that there's
no negative logic anywhere in the diagram. The NAND gates here are all
positive NAND gates. We've just *imagined* negative logic being used, to
help us understand that NAND gates with inverted input signals act as
OR gates. This helped us to follow the "logic" of the NAND network.

What have we learned about digital decisions?

Your head may be spinning, but you've got to admit that you see now what we mean when we say that digital circuits can make decisions. The designer of the system just picks apart the complex decisions that must be made until he gets down to decisions simple enough to be made by gates. He has five kinds of gates to choose from, basically (AND, OR, NOT, NAND, NOR). But the limitations of the particular type of integrated circuit he's working with may cut down the designer's options. Usually, as in our last example, he must make do with NAND, NOR, and NOT gates. But the designer can substitute the correct combination and go ahead with little problem.

In later chapters, we will become more and more familiar with the marvelously adaptable little circuits called gates. We'll become aware of ways to analyze system requirements so as to pick the most efficient combination of available gates. We'll learn about the different "families" of integrated circuits, each based on a particular type of design for gates, and about the particular applications best for each circuit family. And we'll learn how numerous digital building-blocks, subsystems, and complete systems work, based on our understanding of their foundations in those simplest building-blocks of all – digital gates. So you can see that our comprehension of these circuits will carry us far in learning about digital electronics.

Quiz for Chapter 2

1. Which of the following "universal stages" of a system may contain circuits that make decisions?
a. The "sense" and "act" stages.
b. The "decide" stage.
c. The "store" stage.
d. All of the above.

2. Which of the following may be called a "coincidence detector?"
a. AND gate.
b. OR gate.
c. Both of the above.
d. None of the above.

3. Which of the following can act as a "gate in a fence" for information coming through one input, under the control of the other input?
a. A 2-input AND gate.
b. A 2-input OR gate.
c. Both of the above.
d. None of the above.

4. Which is the best way to imagine the internal operation of a positive AND gate?
a. Switches connected in series.
b. Switches connected in parallel.
c. MOS transistors connected in series.
d. MOS transistors connected in parallel.

5. How many inputs can an AND gate or an OR gate have?
a. One.
b. Two.
c. Three.
d. Two or more.

6. The precise, official definitions of the AND, OR, and NOT functions are in terms of:
a. On and off, or high and low voltage.
b. 1 (yes) and 0 (no).
c. Either of the above.
d. None of the above.

7. What do we call a table that shows the *logical* state (1 or 0) of a digital circuit's output for every possible combination of *logical* states in the inputs?
a. Truth table.
b. Function table.
c. Either of the above.
d. Both of the above.

8. What do we call a table that shows the *electrical* state of a digital circuit's output for every possible combination of *electrical* states in the inputs?
a. Truth table.
b. Function table.
c. Either of the above.
d. Both of the above.

9. To decide which gate function an electric circuit with a certain function table performs, you must first decide:
   a. Whether you're using positive or negative electricity.
   b. Whether you're using positive or negative logic.
   c. Whether the "higher," more positive voltage means 1 or 0.
   d. B and C above.

10. How many different truth tables can be made from one function table?
    a. One.
    b. Two.
    c. Three.
    d. Any number.

11. An OR gate's output is 1 when:
    a. All inputs are 1.
    b. All inputs are 0.
    c. Only one input is 1.
    d. One or more inputs are 1.

12. An AND gate's output is 1 only when:
    a. All inputs are 1.
    b. All inputs are 0.
    c. Only one input is 1.
    d. One or more inputs are 1.

13. A positive OR gate's output is "high" when:
    a. All inputs are "high."
    b. All inputs are "low."
    c. Only one input is "high."
    d. One or more inputs are "high."

(Answers in back of the book)

14. A positive AND gate's output is "high" only when:
    a. All inputs are "high."
    b. All inputs are "low."
    c. Only one input is "high."
    d. One or more inputs are "high."

15. Which is the best way to imagine the internal operation of a positive OR gate?
    a. Switches connected in series.
    b. Switches connected in parallel.
    c. MOS transistors in series.
    d. MOS transistors in parallel.

16. The most basic law of semiconductor action is that voltage pressure cannot easily force positive electric charge to flow across the border from:
    a. An n-region to a p-region.
    b. A p-region to an n-region.
    c. An MOS transistor's metal "control plate."
    d. Any of the above.

17. What distinguishes one electrical state from the other in all "real" electronic digital circuits, including MOS circuits?
    a. Current moving or stopped.
    b. Outputs "on" or "off."
    c. Voltage "high" or "low."
    d. Any of the above.

18. A positive AND gate is also a negative:
    a. AND gate.
    b. NAND gate.
    c. NOR gate.
    d. OR gate.

Building Blocks
That Make Decisions

In Chapter 2, we became familiar with the basic building-blocks that make up all digital electronic systems, which are simple little circuits called "gates" (*Figure 3-1*). We noted that designing a digital system is basically a matter of putting together large numbers of gates in the right way. As it turns out, this is a lot like assembling a tinkertoy project from a few kinds of parts. And so, to proceed with finding out how digital systems work, we've got to look into the ways these gates are put together to do various jobs.

**Figure 3-1.
Symbols for Digital
Gates**

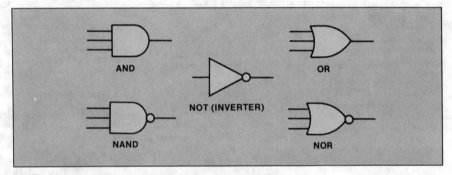

Fortunately for our study of the ways gates are put together, most digital systems are made up of just a few different kinds of *"building blocks" that are themselves made of gates.* So in this chapter and the next, we will become familar with about seven different kinds of these building-blocks made of gates. There are many different varieties of each building-block, but we will learn the basic principles by looking at a typical example of each one. Furthermore, there are other kinds of building-blocks than the ones we'll study. But the ones we'll pick out will resemble the others to a great extent.

How are digital building-blocks classified?

We're going to organize our study of digital building-blocks into two parts. First, in this chapter, we'll look at some units that *don't have any memory in them.* They just make *decisions,* based on the inputs they happen to be receiving at the moment. (See *Figure 3-2.*) These non-memory units are called "combinational" circuits, because for every combination of bits in the various input wires, there's a definite, prearranged combination in the output wires to be decided upon. The output combination is the same every time a particular input combination shows up.

Combinational circuits, which have no memory, have outputs that have a definite prearranged outcome for a given combination of inputs.

Sequential circuits are building blocks with memory capability. Their outputs are functions, not only of present inputs, but of ones received in the past as well.

On the other hand, the building-blocks that *do* contain memory circuits can *store* information derived from *previous* combinations of inputs. So the combination of output bits depends not just on the input signals at the moment, but also on *previous* combinations of bits. These "memory-containing" building-blocks are called "sequential" circuits. This is because the outputs depend on a sequence, or chain of inputs at different times. We'll cover sequential building-blocks in Chapter 4.

**Figure 3-2.
Combinational Digital
Building Block**

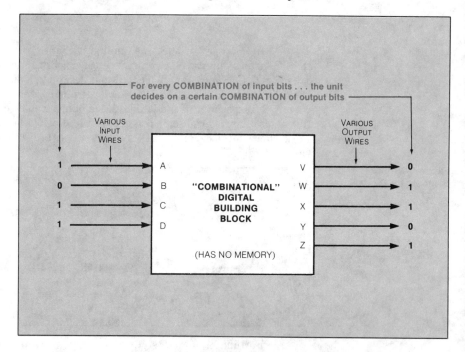

How will we approach an understanding of combinational circuits?

In learning about combinational building-blocks, we could simply look at three or four different gate networks, note what they do, and trace how the gates operate together. But we need to do more than that. We need to start developing your ability to *analyze* combinational networks, so you can look at a network in the future and follow its action by yourself. You may even want to begin designing networks on your own. After all, the parts you'll need are available from most electronic supply companies. And as we said earlier, the parts go together like tinkertoys, provided you stick with the same "family" of circuits. (Some very common families you'll see when we study integrated circuits later are called TTL and MOS.)

With this in mind, then, we'll spend quite a bit of time with the first combinational building-block. We'll look at it from the viewpoint of a designer. This exercise will carry us far in understanding digital systems in general, because combinational analysis is the real heart and core of digital electronics.

What are "code-converter" building-blocks?

A code converter receives one digital code format at its input and translates it at its output.

The first class of decision-making building blocks we'll take up is a type that could be called "code converters." This group includes the part of the keyboard encoder that we studied in Chapter 2 (*Figure 2-8*). It also includes the main part of the segment decoder subsystem in the calculator of Chapter 1 (*Figure 1-5*). A code-conversion building-block simply takes in digital information in one type of code, and puts out the same information in another type of code. For example, in the keyboard encoder (*Figure 2-8*), the incoming code for "5" is 10000010000 in lines N and P and the scan lines. And the output code for "5" is 0101 in lines that signify 8, 4, 2, and 1.

Code converters are a good place to start because the gate arrangements in them are a simple, logical kind you'll find in nearly every other type of combinational building-block. After learning to analyze a code-conversion unit, you'll find it much easier to analyze other networks.

What is a "BCD-to-7-segment decoder"?

Let's narrow our sights now on just one kind of code converter, a "BCD-to-7-segment decoder." This unit would form the main working part of the segment decoder shown back in *Figure 1-5. Figure 3-3* shows what its job is.

**Figure 3-3.
A BCD-to-7-Segment
Decoder**

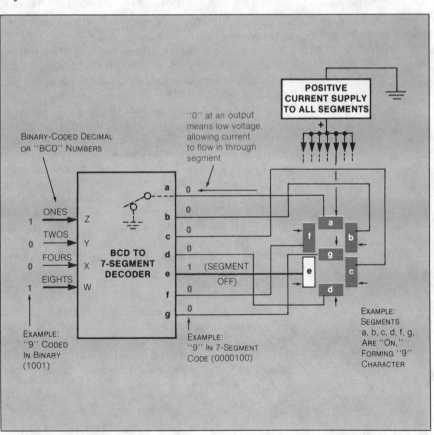

This combinational unit takes in decimal digits coded in binary form as we've seen before. The code is summarized in *Figure 1-18*, and it's called the "binary-coded decimal" or "BCD" code. Notice we've labeled the four inputs W, X, Y, and Z. In *Figure 3-3*, the number 9 is being received.

The seven outputs, labeled a through g, turn on a 7-segment display to show the number being received. Each output is connected to a segment labeled with the same letter. The other terminal of each segment is connected to a source of current (such as a scan line in the calculator). *To turn a segment on, the decoder transmits a zero* in that output. This allows current to flow through the segment and *into the output*. This action is as we described it in Chapter 1, but we're leaving out the decimal point this time to keep it simple. In the figure, "9" is being transmitted in 7-segment code as 0000100, which turns on all segments except "e." (Note that in relating "logical" 1 and 0 to this electrical circuitry, we're using *positive logic*. High voltage means 1 and low voltage means 0.)

Now we've said a number of times that *any* information-processing job can be handled by gates put together the right way. But how in the world would a designer begin to decide how to make a gate network to handle this particular job (or any other, for that matter)?

How do you begin designing a combinational network?

Well, do you remember the truth tables we examined for each gate back in Chapter 2? (*Figures 2-6, 9, 15, 16,* and *17*). The starting point for designing *any* combinational building-block is to make out a truth table for it. The table must show all the possible input combinations, and the output combination that results from each one. *Figure 3-4* shows the truth table for our BCD-to-7-segment decoder.

For your reference, the shape of each numeral character is shown to the left of each horizontal row of the truth table, and a chart of segment labels is provided above. For example, notice in the top row that the numeral 0 is received as 0000 in lines W, X, Y, and Z. And the output code for this numeral is 0000001, which turns on all segments except "g."

Notice that the bottom six input combinations represent the binary numbers ten through fifteen. These combinations have no meaning in BCD code. We will assume that none of them will ever be received by our decoder. However, it will help us later on if we include these "never-received" combinations in the truth table. We write the letter "X" for all the outputs in these cases, to signify "don't care," or "irrelevant."

How does a truth table help design a network?

Okay, how will this truth table help us design our network? Well, any number of gate networks could be designed to obey a given truth table. A designer would want the *best* of all the possible networks, considering what kind of gates he has available to work with. He will probably want the network with the *fewest* gates and the *most economical* gates. This also implies he wants to use gates with the *fewest number of*

A truth table shows the appropriate output response for every possible input condition, including outputs/inputs that have no meaning.

Figure 3-4.
Example of a Truth Table

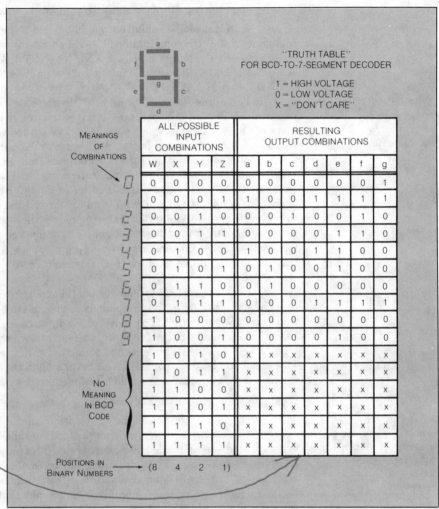

"TRUTH TABLE"
FOR BCD-TO-7-SEGMENT DECODER

1 = HIGH VOLTAGE
0 = LOW VOLTAGE
X = "DON'T CARE"

MEANINGS OF COMBINATIONS

ALL POSSIBLE INPUT COMBINATIONS				RESULTING OUTPUT COMBINATIONS						
W	X	Y	Z	a	b	c	d	e	f	g
0	0	0	0	0	0	0	0	0	0	1
0	0	0	1	1	0	0	1	1	1	1
0	0	1	0	0	0	1	0	0	1	0
0	0	1	1	0	0	0	0	1	1	0
0	1	0	0	1	0	0	1	1	0	0
0	1	0	1	0	1	0	0	1	0	0
0	1	1	0	0	1	0	0	0	0	0
0	1	1	1	0	0	0	1	1	1	1
1	0	0	0	0	0	0	0	0	0	0
1	0	0	1	0	0	0	0	1	0	0
1	0	1	0	X	X	X	X	X	X	X
1	0	1	1	X	X	X	X	X	X	X
1	1	0	0	X	X	X	X	X	X	X
1	1	0	1	X	X	X	X	X	X	X
1	1	1	0	X	X	X	X	X	X	X
1	1	1	1	X	X	X	X	X	X	X

NO MEANING IN BCD CODE

POSITIONS IN BINARY NUMBERS → (8   4   2   1)

Even though these outputs are not used they are included for clarity.

inputs, because each input increases the size and cost of a gate circuit as well as the number of wires that have to be run.

When designing combinational logic circuits, the digital combinations possible from the truth table form the basis for deciding the gate and gate combinations to use for a given output.

However, for the moment, let's concentrate on designing just one particular network that's very straightforward and easy to understand from a *logical* point of view. Later on, we'll get around to simpler, more economical networks. So the question is, how will the truth table help us design this "straightforward" network?

Well, to start with, you design a *separate network for each output wire,* based on the *logic* in the truth table—meaning the *reasoning* that's implied by the ones and zeroes. For example, let's take output "a." In the column under output "a" in the truth table, we have a 1 for input combination 0001 and another 1 for input combination 0100. Now what's the *logic* in this fact, that will lead us to a gate network for output "a"?

How do you pick the "logic" out of a truth table?

Here's the logic behind output "a": "a" is 1 when we receive combination 0001 OR combination 0100. Furthermore, we have combination 0001 when W is NOT 1, AND X is NOT 1, AND Y is NOT 1, AND Z is 1. And we have combination 0100 when W is NOT 1, AND X is 1, AND Y is NOT 1, AND Z is NOT 1.

The words "OR, AND, and NOT" in these logical statements tell us how to connect an OR gate, some AND gates, and some inverters, to make our network for producing output "a." (We'll see how to recognize this network from the statements in a moment.) We can make similar logical statements for each of the other outputs from this building-block, or from *any* combinational network. Statements like these can be drawn from any truth table—statements in terms of the three basic "logical operations" of AND, OR, and NOT. And for each logical statement like this, there's a network of gates that will produce the desired output from the given inputs. It may not be the simplest, or most economical network— but it will work. So the next step in our design process, after writing the truth table, is to understand the truth table in terms of this kind of statement.

However, logical statements like this are very cumbersome to write in plain language, and this plain language doesn't make it very easy to see how the gates are to be connected. So digital designers use a very handy "shorthand" writing for logical statements, which makes the relation between the statements and a gate network very easy to see. This shorthand is such a common, everyday thing in the field of digital electronics that our learning would be incomplete if we left it out.

What's the "shorthand" for writing logical statements?

This logical shorthand was developed by an Englishman named George Boole (rhymes with "pool") long before digital electronics were ever invented, and it's called "Boolean algebra." Its purpose was to provide a neat, simple way to write complicated combinations of "logical statements," which are defined as statements that can be either true or false. "Logic" is a very ancient branch of philosophy that's concerned with the study of logical statements—with proving whether one statement is true if certain other related statements are true or false.

Boolean algebra provided a ready-made way for digital designers to handle the "logical" (true-false) statements that are involved in every binary digital system. (After all, we have learned that 1 means true, and 0 means false.) Because of this involvement with logic, you'll often see digital gates called "logic gates." And a network of gates is sometimes called a "logic network," or a "Boolean network."

AND, OR and NOT Logic gates can be combined to provide the logical functions for outputs derived from the input combinations in the truth table.

Boolean algebra, in the form of mathematical expressions, provides a short, concise form for expressing truth table logic statements.

What are the principles of "Boolean algebra"?

Boolean expressions label inputs and outputs with letter symbols connected, where appropriate, with a multiplication sign for an AND function, addition sign for the OR function, a bar over the letter or an equal sign for the NOT function.

Anyway, *Figure 3-5* summarizes the basic elements of Boolean algebra, as it's related to digital electronics. The idea is that we use the *letter label* of a wire to represent the *logical statement* carried by the wire. And we use a multiplication sign to represent the AND operation, an addition sign to represent the OR operation, and a bar over the letter to represent the NOT (inversion) function. These three operations are the only ones used in Boolean algebra.

For example, as shown in *Figure 3-5*, suppose we have an AND gate with inputs A and B and output X. The way the signal at X is related to those at A and B is expressed by this logical statement: "X is 1 if A is 1 AND B is 1." Using Boolean algebra, this statement would be written simply as $X = A \cdot B$, which you would read, "X equals A AND B." (You could just as well write $X = AB$, which is another way of writing the AND function.)

**Figure 3-5.
Summary of Boolean
Algebra**

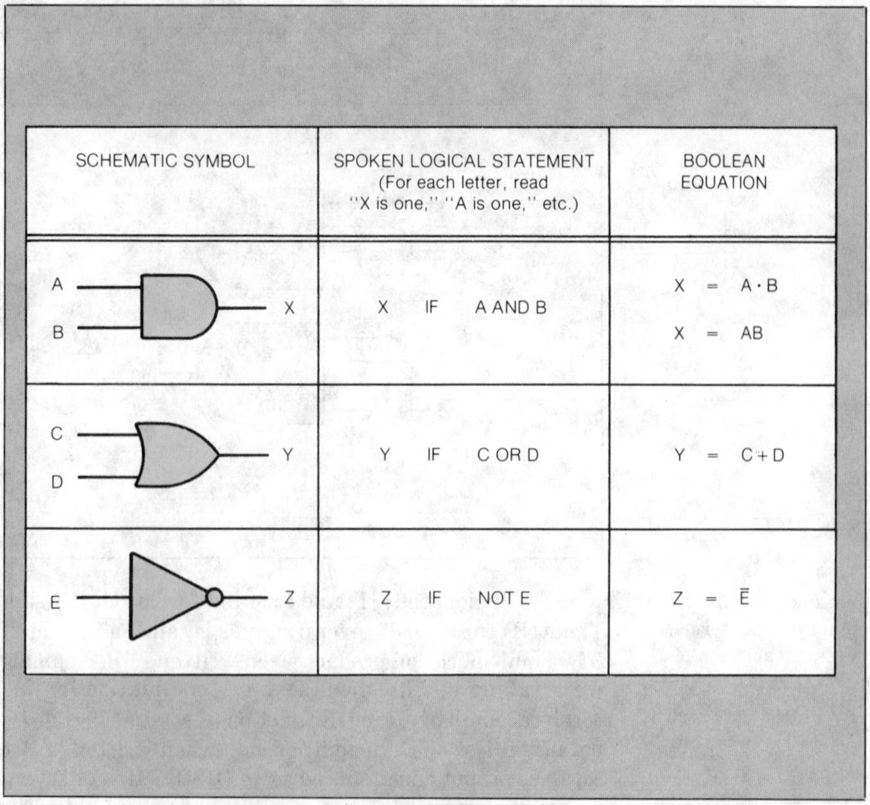

SCHEMATIC SYMBOL	SPOKEN LOGICAL STATEMENT (For each letter, read "X is one," "A is one," etc.)	BOOLEAN EQUATION
A, B → X	X IF A AND B	$X = A \cdot B$ $X = AB$
C, D → Y	Y IF C OR D	$Y = C + D$
E → Z	Z IF NOT E	$Z = \overline{E}$

Similarly, if Y is the OR function of C and D, you would write $Y = C + D$. And if Z is the inverse of E, you would write $Z = \overline{E}$.

This little bit of Boolean algebra is enough to help us out a lot in designing our decoder. So let's get back to that job, by looking at *Figure 3-6*.

How can Boolean algebra show us a gate network?

Product forms in logic
equations, written in
shorthand form as $P_1 P_4$,
are AND functions; sum
terms are written as $P_1 +$
P_4 and are OR functions.

For your reference, in the upper left area of *Figure 3-6* is
reproduced the part of the truth table that tells us how to design our gate
network for output "a." To the right of this little truth table is the Boolean
equation for "a," which is just a "shorthand" version of the long and
cumbersome statements we made earlier: $a = \overline{W} \cdot \overline{X} \cdot \overline{Y} \cdot Z + \overline{W} \cdot X \cdot \overline{Y} \cdot \overline{Z}$. And
down below is the gate network derived from this equation. The inverters
produce the NOT functions we need of W, X, Y, and Z. The AND gates
produce the two AND functions, which we are labeling P_1 and P_4, and the
OR gate produces the OR functions, which we can write as $P_1 + P_4$. It's as
simple as that.

**Figure 3-6.
A Sum-of-Products Gate
Network**

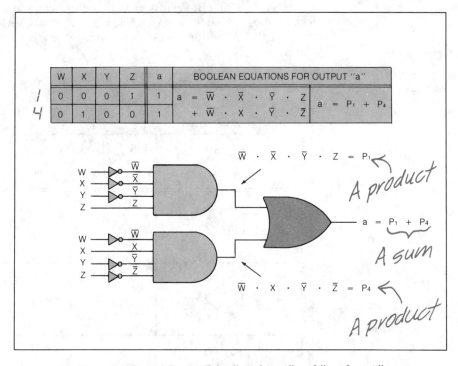

Final Boolean expressions
contain product terms,
sum terms and sum-of-
products terms, the latter
is for AND gates feeding
OR gates.

Incidentally, P_1 and P_4 stand for "product 1" and "product 4."
The AND function of several signals is called a "product" because it *looks
like* a multiplication product when written in Boolean algebra. The signal
we're calling P_1 tells when the input combination for the number 1 is
received. Similarly, the OR function of several signals is called a "sum"
because it *looks like* an addition sum when written in Boolean algebra. And
so, since output a is produced as an OR function of several AND functions,
we say it's produced as a *sum-of-products* . Any network consisting of
AND gates feeding OR gates is called a "sum-of-products" network.

If you like, you can write a sum-of-products equation for the rest
of the seven outputs of our decoder. But to save time for now, let's go right
to the complete network for the entire building-block, shown in *Figure 3-7*.

**Figure 3-7.
Sum-of-Products BCD-
to-7-Segment Decoder-
Driver**

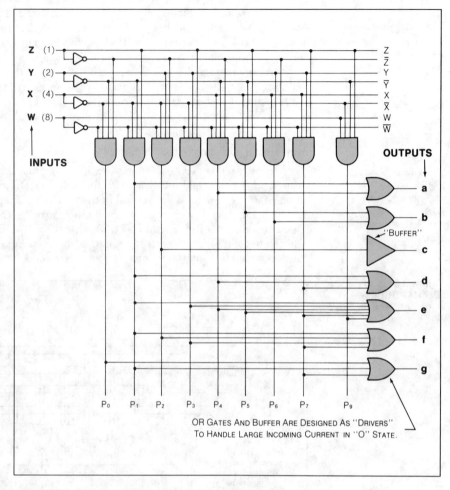

How does the entire decoder network operate?

In designing a decoder, each output can be described with a Boolean equation.

In *Figure 3-7*, notice that we have an AND gate for each of the nine input combinations that we want to produce 1 in any output. Notice that we don't have an AND gate for input combination "8." If you'll look back at the truth table (*Figure 3-4*), you'll see that there are no "1" outputs for input combination P_8. Therefore none of the Boolean output expressions include P_8, and we don't need an AND gate here.

Inputs to the AND gates come from either the "true" inputs W, X, Y, and Z, or from the "complement" inputs $\overline{W}$, $\overline{X}$, $\overline{Y}$, and $\overline{Z}$, which are produced by four inverters at the upper left in *Figure 3-7*. The output from each AND gate is labeled below with the "product" symbol for the AND function the gate produces. In Boolean form, $P_0 = \overline{W}\cdot\overline{X}\cdot\overline{Y}\cdot\overline{Z}$, $P_1 = \overline{W}\cdot\overline{X}\cdot\overline{Y}\cdot Z$, and so forth. Since the OR gate for output "a" is fed from lines P_1 and P_4, "a" is the signal we discussed earlier: $\overline{W}\cdot\overline{X}\cdot\overline{Y}\cdot Z + \overline{W}\cdot X\cdot\overline{Y}\cdot\overline{Z}$. Similarly, all the other output functions are taken directly from the truth table.

Buffers are added between
logic gate outputs and the
output line to provide additional capability to drive
the load on the output.

Now output function "c" is just P_2, so this output could
conceivably be connected directly to product line P_2. But to avoid
burdening the AND gate for P_2 with the heavy current from the display,
we provide the "buffer" shown at output "c." A buffer acts as two inverters
in series, to repeat and strengthen a digital signal. In this particular
building-block, the buffer and OR gates are all specially made to handle the
large segment current when they're in the 0 state. They're called "segment
drivers." Actually, the proper name for the entire building-block would be
"BCD-to-7-segment decoder- *driver.* "

How would you make a real decoder?

In actual practice, inte-
grated circuits are de-
signed using the type gate
that can be repeated and
that is most economical for
a given manufacturing
process.

A real decoder would in most cases differ from this design in two
respects. First, it would substitute (for the AND and OR gates) whatever
kinds of gates were most economical in the particular integrated-circuit
family being used. For example, as shown in *Figure 3-8*, the decoder in the
Texas Instruments TTL circuit SN74143 substitutes NAND gates as we
learned to do in *Figure 2-18.* And second, as shown in *Figure 3-8*, the real
decoder would *leave off* the $\overline{W}$ inputs to the AND gates for P_2 through P_7,

**Figure 3-8.
SN74143 BCD-to-7-
Segment Decoder-Driver**

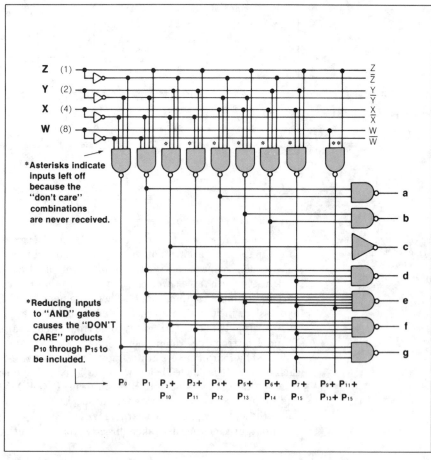

and the $\overline{X}$ and $\overline{Y}$ inputs to the AND gate for P_9. These inputs would only serve to distinguish between these six input combinations and the six "never received" combinations at the bottom of the truth table (which we can call P_{10} through P_{15}). So by taking note of the "don't care" conditions when we make up a truth table, we can often simplify our designs.

(Incidentally, the manufacturers of integrated circuits sell or give away catalogs and data sheets that show the logical design of many digital ICs. You're learning to understand these designs, so you may want to obtain some of these publications to advance your comprehension of digital electronics.)

So that finishes our introduction to the "code-converter" class of combinational building-blocks. All units in this category work more or less like this one—including the part of the keyboard encoder we studied in Chapter 2.

But more than just learning about one type of building-block, you've become acquainted with the use of a truth table and Boolean algebra in analyzing *any* combinational network. This understanding will serve you well as we proceed through our study of digital systems.

How do "data routing" units resemble decoders?

In a data selector, by changing the digital code received at the selection control lines, different input sources may be routed to the output line.

The second class of combinational building-blocks we'll take up could be called "data routing units." Networks in this group actually *route* data (meaning information being processed) from various sources to various destinations. First, *Figure 3-9* shows a network called a "data selector," which as you can see looks a lot like our decoder in *Figure 3-7*, only simpler. As suggested by the "switch" drawing to the right, this unit acts as a sort of "switchboard" with four input wires (A, B, C, D) and one output wire (E). The network allows you to connect any one of the inputs to the output, for the purpose of transmitting digital information. An input is *selected* by means of the two "input selection control" wires, W and X.

**Figure 3-9.
A Data Selector**

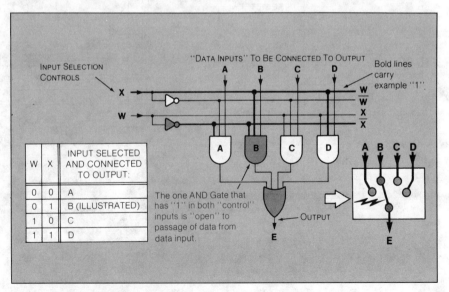

W	X	INPUT SELECTED AND CONNECTED TO OUTPUT:
0	0	A
0	1	B (ILLUSTRATED)
1	0	C
1	1	D

The one AND Gate that has "1" in both "control" inputs is "open" to passage of data from data input.

For example, the bold lines in *Figure 3-9* show how the input combination 01 in W and X select input B. AND gate "B" is the only one with "1" in both of its "control" inputs. As we noted in Chapter 2, this gate's output will now "follow" input B as it carries ones and zeroes, one after the other. And since all other AND gate outputs are "0," the OR gate output E follows this one. So input B is literally "gated through" to the output.

For a *four-input* data selector like this one, we need *two* selection-control inputs, because this gives us four different combinations of control inputs. Three control inputs would let us handle *eight* data inputs (each with its own AND gate feeding the OR gate), four could handle 16, etc.

How would a data selector be used?

On the other hand, you often see *two-input* data selectors—and they need just *one* control line. *Figure 3-10* shows an example as it might apply to our example calculator system. As shown below in this figure, we want to be able to route four-bit numbers (A and B) either "straight through" from the display and operand registers to inputs X and Y of the adder-subtracter, or to make the numbers "cross over" to the opposite inputs—A to Y and B to X. This will let us either subtract B from A, or subtract A from B.

By selecting appropriate inputs with a two-input data selector, input bits can be routed to an adder-subtractor. *Figure 3-10* only shows one bit each of two 4-bit inputs.

**Figure 3-10.
2-Input Data Selectors in
Adder-Subtracter**

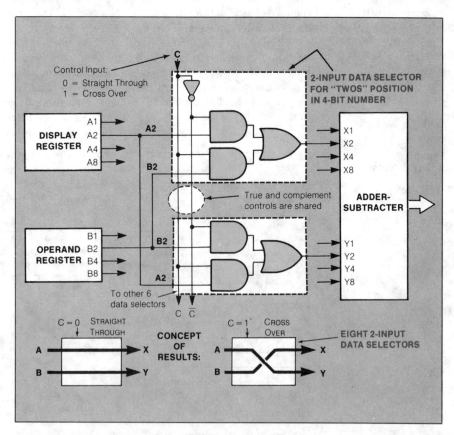

To do this, we use a *two-input data selector* on each of the eight input wires to the adder-subtracter. (That's eight data selectors in all—but only two are shown.) We need only one control input, labelled "C". A "0" sends the numbers straight through, and "1" makes them cross over. So as you can see, this concept of data selection can be very versatile.

**Figure 3-11.
A Demultiplexer**

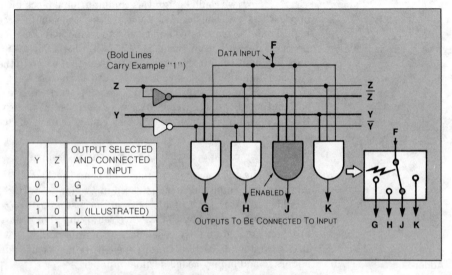

How do you "demultiplex" an input to one of several outputs?

Figure 3-11 shows a very similar kind of "data-routing" building-block called a "demultiplexer." This one has the *opposite* job from the data selector—it routes data from *one input* (labelled F) to any one of *several outputs* (G, H, J, and K). As you can see, the selection process works the same way as for the data selector.

What kind of building-block adds numbers?

Let's look at one more combinational building-block now—a four-bit binary adder. This unit's job is illustrated in *Figure 3-12* . It takes two four-bit binary numbers A and B, and produces a *five-bit* binary sum of A plus B. (Not the Boolean "OR" function this time, but an arithmetic sum.)

**Figure 3-12.
A 4-bit Binary Adder**

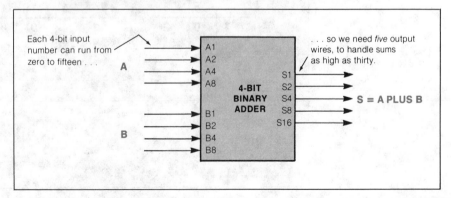

Designing a 4-bit binary adder requires a different approach than using the sum-of-products network that is used in the decoder-driver and the data-selector.

As you can tell from the label "S16," this fifth output wire has a value of 16, compared to values of 8, 4, 2, and 1 for the other four. Each input number can run from zero (0000) to fifteen (1111). So the sum can run from zero (00000) to thirty (11110). If the "sum" output had only four wires like the inputs, it could only go up to fifteen.

Now we could design this combinational block by using the same approach as with the decoder—by designing the sum-of-products version right from the truth table and trying to simplify it. But the sum-of-products network would need *256 eight-input AND gates* to handle all the possible input combinations in eight wires (that's two to the eighth power—two multiplied by itself seven times). And the output OR gates would have to handle as many as *128 inputs* (which isn't practical—we'd have to handle each output OR function in two or more "stages," say, with sixteen 8-input OR gates feeding a seventeenth). So we have to back off, scratch our heads, and find another way.

What's a simple approach to a binary adder design?

The answer comes from considering the "logic" (the reasoning) of adding binary numbers, as illustrated in *Figure 3-13* . Here, we see 44 and 58 added to get the sum of 102, both in the customary *decimal* way and in the *binary* way. (And by the way, we're talking about "pure" binary numbers here, not binary-coded decimal numbers. That's why "44" is 101100, and not 0100 0100. Later on, we'll see how to use pure-binary adders to add binary-coded decimal numbers.)

**Figure 3-13.
Binary Numbers Are
Added Like Decimal
Numbers**

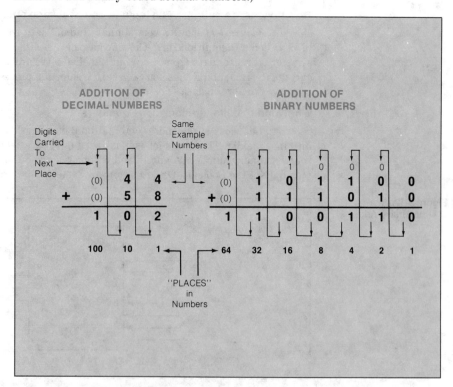

Binary numbers are added in a manner similar to adding the more familiar decimal numbers.

Binary numbers are added by the same process you've always used to add *decimal* numbers. You just line up the two numbers, so that the "ones" place in one number is over the "ones" place in the other, and so on. Then you start at the right end (the "least significant" end, where the smallest "place" is), and you add *one pair of digits at a time*. When you get a two-digit result from one of these steps, you "carry" the extra "one" to the next step to the left, and add it in with these two digits.

Aha! Is a light coming on in your head? To add our two 4-bit numbers, all we need is four *one-bit adders*, right? That way, we can handle this nasty combinational job in *four simpler stages*! (Remember, the magic words for digital systems are, "Break up the big jobs into little jobs!")

How can we use one-bit adders to add multi-bit numbers?

Figure 3-14 shows how we do this. We use four identical little building-blocks called "1-bit full adders." Each one adds two bits called A and B, plus a third bit carried in from adding less-significant digits. Each 1-bit adder produces a 1-bit "sum" output S, and a carry signal to the next more significant digit of addition. Obviously, we can chain together as many of these little adders as we might want to, to handle input numbers with as many bits as we like.

Multiple bit binary adders can be formed by connecting one-bit adders in a chain. Each adder adds two binary bits together and has the capability of transferring a carry to the next adder.

However, for the 4-bit adder of *Figure 3-12*, we only need the four 1-bit adders shown in *Figure 3-14*—we don't have any "preceding" or "following" stages of addition. So we just supply a continuous "0" to the "carry-in" input N to the least-significant adder. (If we wanted to, we could make this particular 1-bit adder on the right a "half-adder," which wouldn't have a "carry-in" input.) The layout in *Figure 3-14* is typical of many "real" 4-bit adders, including the Texas Instruments SN74LS83. You'll find this IC in *The TTL Data Book*, sold by Texas Instruments.

Figure 3-14.
Using Binary Full Adders

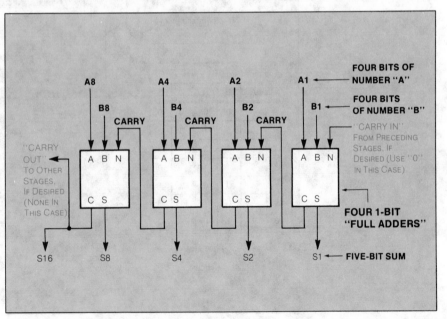

How does a one-bit full adder work?

The one-bit full adder has an incoming carry signal (if any) from a previous adder, and one bit signals from each of the two numbers to be added.

Figure 3-15 shows the truth table for a 1-bit full adder—including all eight possible combinations of inputs A, B, and N, and the desired combinations in outputs C and S. The network shown in this figure is just about as simple a design as you can get, if you're working only with AND gates, OR gates, and inverters. (The Texas Instruments SN74LS83 we mentioned modifies this a little bit by substituting other kinds of gates—in particular, an "AND-OR-invert" circuit that's very economical in the TTL family, which we will see in the chapter on integrated circuits.)

The network drawing in *Figure 3-15* is covered with road-signs like a highway, to help you find your way through it if you like. For reference, each input combination in the truth table is identified by a number from zero through seven, corresponding to the binary number formed by the combination. The AND function of an input combination is designated by a "P" with the same number subscripted, so that $P_1 = A \cdot \overline{B} \cdot N$ and $P_7 = A \cdot B \cdot N$, for instance.

**Figure 3-15.
Truth Table for a 1-bit Full Adder**

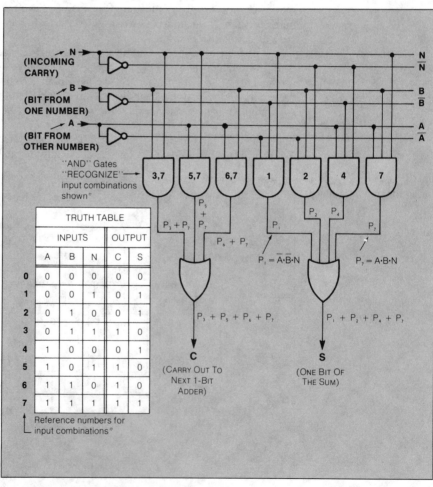

What have we learned about combinational building-blocks?

That just about covers the main, basic combinational building-blocks you're likely to run into in digital systems: various kinds of code-conversion units, data-routing units, and adders. There are endless varieties and combinations of these types of networks, but the concepts by which they all work are pretty much as we have seen in this chapter.

The important, unifying thread that runs through all these "decision-making" building-blocks is this: for every combination of signals at the inputs to a unit, there's a certain, prearranged combination of signals at the outputs. Furthermore, these input and output combinations can be summarized neatly in a truth table. Boolean algebra provides a convenient "shorthand" method of writing the logic (the reasoning) that shows how to produce each output signal. This method uses symbols for statements and for the three basic logic operations.

We've noted that the gate network derived directly from a truth table is not necessarily the most economical. The network can possibly be simplified by taking advantages of any "don't care" conditions in the truth table. Different gates can be substituted according to principles we glimpsed in Chapter 2, to make best use of gates available in the particular integrated-circuit family being used. Also, as we saw in the adder, a combinational job can be handled in several steps if necessary to simplify the network.

So as we said at the beginning of the chapter, we've done considerably more than just look over some building-blocks. We've deepened our comprehension of the principles of digital electronics by looking at things from the designer's viewpoint. All this background understanding will serve you well as we continue our study of digital building-blocks in the next chapter with *sequential* networks, and as we proceed later on to a comprehension of how entire systems are put together from building-blocks.

1. Which kind of building-block
   can store information?
   a. Combinational
   b. Sequential
   c. Neither
   d. Both

2. Where do we get the name,
   "combinational?"
   a. Such building-blocks are
      combinations of gates.
   b. They combine several
      inputs to create outputs.
   c. For every combination of
      input signals, there's just
      one combination of output
      signals.
   d. A and C above.

3. What information is shown in
   a truth table?
   a. Whether the unit tells the
      truth or not.
   b. Output combination for all
      input combinations.
   c. Just the outputs that are
      "true" or "1."
   d. None of the above.

4. What does an "x" mean in
   outputs in a truth table?
   a. The output is neither "1"
      nor "0."
   b. We don't care what this
      output is.
   c. The letter "x" is formed in
      the display.
   d. A and B above.

5. What does "sum-of-products"
   mean in Boolean algebra?
   a. The OR function of several
      AND functions.
   b. The AND function of
      several OR functions.

(Answers in back of the book)

c. The OR function of several
   OR functions.
d. The AND function of
   several AND functions.

6. How does a "real"
   building-block design usually
   differ from the
   sum-of-products design?
   a. There are usually fewer
      gates and fewer inputs to
      gates.
   b. Other kinds of gates may
      be substituted for AND
      gates and OR gates.
   c. It's usually more
      economical.
   d. All of the above.

7. What is the largest number of
   data inputs a data selector can
   have if it has two control
   inputs?
   a. Two
   b. Four
   c. Eight
   d. Any number.

8. In a demultiplexer or data
   selector, the AND gate that's
   open for passage of data from
   one input is the one whose
   inputs are:
   a. Connected to the "true"
      input lines.
   b. Connected to the
      "complement" input lines.
   c. All "1."
   d. All "0."

9. Which building-block studied
   best illustrates the principle
   of dividing big processing jobs
   into a sequence of little ones?
   a. The decoder
   b. The data selector
   c. The demultiplexer
   d. The 4-bit adder

Building Blocks with Memory

In Chapter 3, we found out about "combinational" building blocks. These are digital units whose outputs depend only on the inputs being received at the present moment. For every combination of input signals, there's a certain prearranged combination of output signals.

And now we're going to move on and study the other category of building-block, the "sequential" type. These units can *store* information inside them. So the outputs do *not* depend only on the present inputs, but also on input signals received in the *past* — that is, on a *sequence* of inputs. As we will see, this "memory" capability calls for a set of additional rules and thus adds complexity to the building-block and the system, in terms of both design and operation.

In a later chapter, we will be looking at *subsystems* called "mass memories" that store relatively large quantities of information. But right now, we're talking about smaller units that are regarded as *building-blocks*. When we study mass memories later on, we will see that there are several different techniques used for storing digital information. But in this chapter, we'll limit ourselves to storage elements of just one very important type. These are circuits called "flip-flops," which are made of relatively simple assemblies of gates. Though gates are *combinational* units, they can be hooked up in a way that actually *stores* information. So let's take flip-flops as our starting point.

What's a simple example of a "flip-flop?"

The gated latch has an input gating network coupled to a basic latch. Inputs are gated through the input network to control the condition of the latch only when the correct control signal is applied to the gating network.

To unravel the mystery of how a *sequential* circuit can be made of *combinational* units (gates), let's begin with a very simple type of flip-flop. Let's go back to the "latch" we looked at in general terms back in Chapter 1 (*Figure 1-19*), and see how such a circuit can be made from gates.

To distinguish this circuit from a smaller one inside it that's also called a "latch," let's call the larger circuit a "gated latch." One way to make it out of gates is shown in *Figure 4-1*.

Externally, this circuit is a direct copy of the latch in *Figure 1-19*. It passes signals from the "data input" D straight through to the "true output" Q whenever the "gating input" G is 1. (The "complement output" $\overline{Q}$, pronounced "Q-bar," is the *complement or opposite* of Q, and is produced as part of the internal operation of the circuit.) When G is changed to 0, the output is "latched," or held, in its present state, *regardless of changes in D*, until G goes back to 1. Thus, one *bit* of data is *stored* while G is 0. ("Data" means information being *processed*, as opposed to *control* information.)

Incidentally, you should note that in most kinds of flip-flops, a signal that causes storage to occur at a given time (such as the gating signal G) is called the "clock" signal. This is because such a signal tells the *time* at which data is to be stored and released. Clock signals *synchronize* the changing of data in various parts of the system — meaning make it happen at the same time. However, "gating" is a clearer word in this case.

Figure 4-1.
A Flip-Flop Called a
Gated Latch

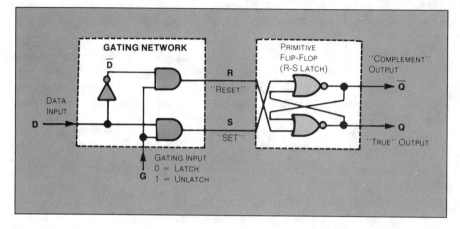

How does the gated latch work?

When G = 0, no input levels can set the R-S Flip-Flop. When G = 1, the levels on D and D̄ determine the state of the R-S flip-flop according to the truth table of *Figure 4-7*.

The two AND gates and inverter in the outlined box at the left in *Figure 4-1* are called the "gating network." This network serves to pass signal D and its complement D̄ through to lines R and S whenever G is 1. So whenever the circuit is "unlatched" like this, R and S are complements of each other. But when G changes to 0, the gating network makes both R and S go to 0. That is, whichever one was at "1" changes to 0. And the network *holds* R and S at 0 while G is 0. (You can easily verify this operation of the gating network for yourself.)

Now looking over to the right, the two NOR gates in the outlined box are actually a basic kind of flip-flop called an "R-S latch." This is the *sequential* part of the gated latch, and the part where the bit of data is actually stored. We won't trace through its operation. (You can do that for yourself, keeping in mind that the output of a NOR gate is 0 whenever at least one input is 1.) We'll just point out that what *makes* it sequential is the way the output of each NOR gate is "fed back" to one input of the other NOR gate. We say that the gates are "cross-coupled."

As a result of this cross-coupling, the true output Q is the *same as S so long as R and S are complements of each other*. This makes Q the same as D when G is 1. But *changing R or S to 0* (whichever one is 1) *does not affect the outputs*. (This happens when G changes to 0.) The changing signal (R or S) has no effect on the NOR gate it feeds because the other input to this NOR gate is 1, fed back from an output. Thus, cross-coupling makes the NOR gates *keep each other in the same state* when one input (R or S) *changes* to 0 while the other *remains* at 0. This stores a bit of information in spite of the change.

Where do the names used in this circuit come from?

The name "R-S" for the NOR-gate latch comes from the customary names for its two inputs, which are "reset" and "set." A momentary 1 in the S line while R is 0 will "set" the true output to 1. And a momentary 1 in the R line while S is 0 will "reset" the true output to 0.

The name "gated" latch for the entire circuit comes from the fact that additional *gating* is provided on the inputs to the R-S latch. This gating is sometimes called a "steering network," because it seems to "steer" the R-S latch along the way it's supposed to go, in response to the D and G inputs.

And finally, the general name "flip-flop" for this type of circuit (both the entire gated latch and the R-S latch alone) is a beautiful way to describe the way such a circuit "flips" to one state and "flops" back again. In this respect, a flip-flop acts somewhat like a toggle switch on the wall for a lamp. It will stay in one position until something comes along and "flips" it. (Thus, a light switch is a digital memory device!)

How are flip-flops used in a "parallel register"?

A little later, we'll get to know several other kinds of flip-flops. They all serve to hold and release one bit of information in response to various kinds of input signals. But first, let's see how the gated latch from *Figure 4-1* could be used in a very important type of sequential building-block called a "parallel register." *Any* flip-flop could be used in this general fashion, but we'll illustrate the idea with the gated latch.

Figure 4-2 shows a digital system (a digital voltmeter) using twenty gated latches grouped as a parallel register. The latches could be bought in groups of four as the Texas Instruments "TTL" integrated circuit SN7475, called a "four-bit latch." If this IC is used with positive logic, its four gated latches are identical to the one in *Figure 4-1*.

Flip-flop describes the way the R-S latch flips between its two stable states based on signals applied at the R-S inputs. When $S=1$ and $R=0$, the Q is set to a 1; When $S=0$ and $R=1$, Q is reset to a 0.

A parallel register is a very important sequential building block that consists of a group of flip-flops with common gating and/or clock signals. Outputs can be transmitted all at the same time (parallel data).

**Figure 4-2.
Gated Latches Used in
Digital Voltmeter**

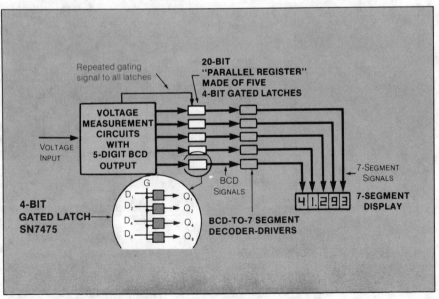

A parallel register consists of a group of flip-flops with a common gating (or "clock") signal. It's used to store information transmitted and received all together, rather than one bit at a time. A group of bits transmitted and received at the same time like this is called "parallel data."

How might a digital voltmeter use a parallel register?

In the highly simplified digital voltmeter system shown in *Figure 4-2*, the parallel data consists of voltage readings in digital form. This data is transmitted from the measuring circuitry at the left in the form of five decimal digits, each coded as four binary bits. (As we know, this method is called "binary-coded decimal," or BCD.) Further toward the right, each decimal digit is decoded into 7-segment code by five BCD-to-7-segment decoder-drivers of the type we studied in Chapter 3. These units drive the five-digit 7-segment display.

However, the measured voltage may be *changing* while it's being read. During some measurements, the last two or three digits to the right are likely to be just a blur of changing numbers. So we have the latches ahead of the decoder-drivers, to *hold* the display constant for a moment every few seconds, so we can read it. The latching is controlled by a regularly repeated gating signal to all the latches. During the intervals when the latches are "unlatched," we can see how fast the voltage is changing, by watching the changing digits.

**Figure 4-3.
A Parallel Register vs a
Shift Register**

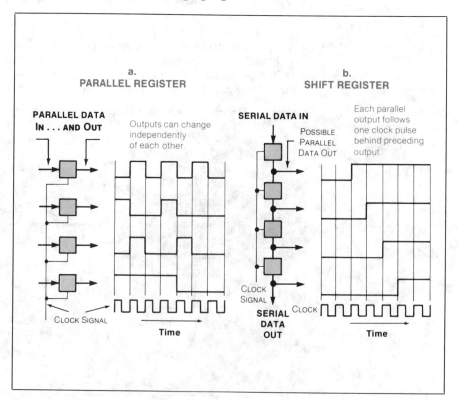

**a.
PARALLEL REGISTER**

PARALLEL DATA
IN . . . AND OUT

Outputs can change
independently
of each other.

CLOCK SIGNAL

Time

**b.
SHIFT REGISTER**

SERIAL DATA IN

POSSIBLE
PARALLEL
DATA OUT

Each parallel
output follows
one clock pulse
behind preceding
output.

CLOCK
SIGNAL

SERIAL
DATA
OUT

CLOCK

Time

How do parallel registers differ from "shift registers"?

In a shift register, also a very important sequential building block, bits shift sequentially, one place at a time all at the same time, under control of a timing signal. Such data movement is called serial data.

This, then, is the idea of how several gated latches (or any other kind of flip-flop) can be used as a *parallel register*. Parallel bits are gated in and out of all the flip-flops *synchronously* (at the same time), as shown in *Figure 4-3a*. (Note that in *Figure 4-3*, we're now using the more general term "clock" signal for what we've been calling the "gating" signal in the case of the gated latch.)

Now the only other way to transmit groups of bits is in the same wire *one at a time*. We call this "serial" transfer, as opposed to parallel transfer. Storing a row of bits coming along one at a time requires another type of register called a "shift register," shown in *Figure 4-3b*. A shift register consists of several flip-flops in a row, with the "true" output of each flip-flop feeding the input of another. Every time the clock signal goes to 1 and back to 0, all the stored bits *shift* from one flip-flop to the next, like buckets being handed from one man to the next in a "bucket-brigade" line. Shift registers are another very important type of digital building-block. In a later chapter, we'll see how they can be used in the three "number registers" in our example calculator from Chapter 1. But for now, let's look at a simpler example illustrating a typical use of shift registers.

What's a typical application of a shift register?

One common use of a shift register is to convert data from serial to parallel or parallel to serial.

Note in *Figure 4-3* that while a shift register is holding a group of bits that were *received serially*, the data can be *read out in parallel fashion* from all flip-flops at the same time (a selected moment when the register is not making a shift). So one obvious use of a shift register is to *convert* data from serial to parallel form.

To illustrate how a shift register can make such a conversion, let's make up a simple example, as shown in *Figure 4-4*. Here, we're supposing that because of space limitations on a calculator IC chip, we only have *one conductor* between the keyboard encoder and the display register. The two four-bit shift registers shown would let us transmit *four-bit* digits from the encoder to the display register *in series*, one bit at a time.

**Figure 4-4.
Parallel Data to Serial
Data and Back Again**

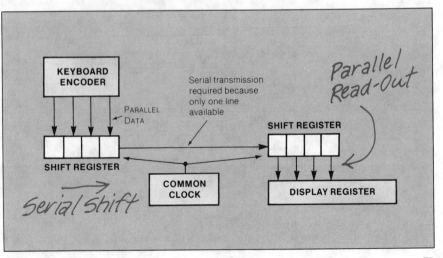

First, the four bits of a decimal digit are loaded *in parallel* into the left-hand shift register. (We'll see how this is done a little later.) Then the two shift registers are shifted four steps by a common clock signal, so that the four stored bits shift out of the left-hand register and into the one on the right. There, the bits are read out in parallel again to the display register. So you can see how this shift register on the right converts data from serial form to parallel form.

How is this system synchronized?

Notice in *Figure 4-4* how important it is for the parts on the left to be *synchronized* with the parts on the right, to be sure the registers shift in step together, and to get the parallel data into and out of the shift registers on the right schedule. The simplest way to synchronize everything is by a "common" clock or timing signal (one supplied in common to all units), as shown in *Figure 4-4*.

You'll see this element of *timing* again and again in digital systems. This example reminds us again why the "storage control" signal to a flip-flop is called a "clock" signal.

While we're thinking about timing, there's an important fact to consider. In shift registers and other similar situations, data must be transferred from one flip-flop to another, with both flip-flops triggered by the *same clock signal*. Shifting bits can be a tricky business when all the flip-flop outputs change at exactly the same time. It requires flip-flops that store their input signal *before* they change their output signal.

How can a flip-flop store before changing?

The most common way to make a flip-flop that can do this is to build two little flip-flops into one bigger one. A "two-step" flip-flop like this is called a "master-slave" type. *Figure 4-5* shows one kind of master-slave flip-flop.

> All activities performed in the register shift operation are timed by the same clock signal. This ensures that all actions are coordinated and occur in the order intended.

**Figure 4-5.
An R-S Master-Slave
Flip-Flop**

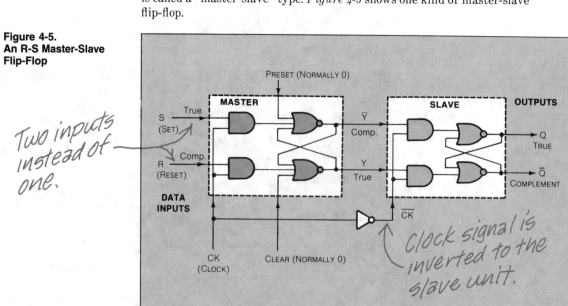

This kind is called an "R-S master-slave flip-flop." It's made of two identical units that we are calling "clocked R-S flip-flops" (shown outlined), one called the master and one called the slave. Each clocked R-S flip-flop is simply a gated latch as in *Figure 4-1*, but with two data inputs rather than one input and an inverter. The "preset" and "clear" inputs are additional features that we'll come to in a moment.

Shift registers require flip-flops that store their input signals before they change their output signals. This is accomplished with a master-slave flip-flop.

Note that the clock signal $\overline{CK}$ to the slave unit is the *inverse* of the main clock signal CK, which is supplied to the master. This gives us the "store-before-changing" action we want. Let's suppose that inputs S and R are connected to the Q and $\overline{Q}$ outputs of another identical flip-flop controlled by the same clock signal. When the main clock changes from 0 to 1, the slave holds one bit while the master accepts a new bit. Then when the main clock goes to 0 again, the master holds the new bit while the slave releases the old one and begins transmitting the new information straight through from the master.

Thus, when this flip-flop is used with others in a shift register, all the flip-flops "get ready to shift" when the main clock goes to 1, which we call the "leading" edge of the clock pulse. And the flip-flops complete the shift at the "trailing" edge of the clock pulse, when CK goes back to 0 again.

So that's how the "master-slave" idea allows bits to be shifted from one flip-flop to another by the same clock pulse going to all flip-flops. We can string together as many flip-flops of this kind as we need in the form of a shift register.

What do the "preset" and "clear" inputs do?

Preset is an input signal that sets a flip-flop to a 1. Clear is an input signal that "clears out" all data by setting the flip-flop to a 0. They both must occur at particular times.

Now what about the "extra features" in *Figure 4-5* that we passed over earlier — the preset and clear inputs? These two inputs are normally kept at 0 and have no effect on the operation of the flip-flop. Putting a momentary "1" pulse into the "preset" input while the clock is 0 (between clock pulses) will store a 1 in the flip-flop. Similarly, a short "1" pulse at the "clear" input between clock pulses will store a 0.

Not all flip-flops have these features. They can be particularly handy in a register, where some systems need to "clear out" all old data and leave only zeroes stored, or to "preset" the register to all ones. (Back in *Figure 1-9*, we saw that our calculator has to be able to clear its registers.) In such a case, the same preset and clear signals would go to all the flip-flops in the register.

How can you load parallel data using "preset" and "clear"?

Another way in which the preset and clear inputs can be useful is in putting *parallel* data into a shift register. We saw a typical need for a parallel-entry shift register in *Figure 4-4*. There, the keyboard encoder puts four parallel bits at once into a shift register. Then the bits are shifted serially through the single line.

Figure 4-6 shows one way to use the preset and clear inputs for parallel data entry. (This method is used in the Texas Instruments SN74165 shift register.) Here we see two stages of a shift register made of R-S master-slave flip-flops as in *Figure 4-5*. Each flip-flop is shown as a customary symbol. (As indicated in *Figure 4-7*, the peculiar circle and triangle symbols at the clock input of each flip-flop are to indicate that the outputs of the flip-flop change when the clock goes from 1 to 0.)

**Figure 4-6.
Loading Parallel Data
Using Preset and Clear**

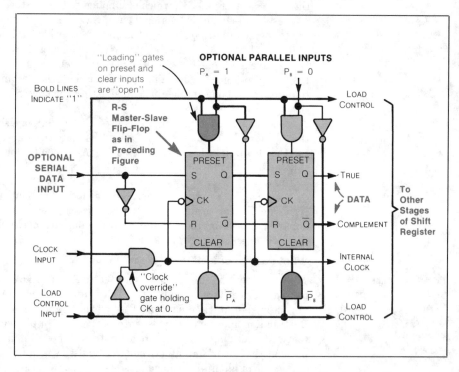

The load control provides a means of overriding the normal clock signal and allowing the preset and clear inputs to function. When the load control line is not active, the register functions as a normal shift register.

Looking again at *Figure 4-6*, note that when the "load-control" input is 0, the circuit functions as a plain shift register, receiving data *serially*. But when the load-control input is 1 as shown in *Figure 4-6*, *parallel* data is entered from the inputs above. The load-control signal makes the "clock-override" gate hold the internal clock signals at 0 (which is required for presetting and clearing these particular flip-flops). It also "opens" the "loading" gates to the parallel data. A "1" at a parallel input such as we see at P_A will operate the "preset" input of a particular flip-flop. And a "0" such as at P_B will be inverted to 1 (as shown for P_B) and operate the "clear" input. As soon as the load-control input is returned to 0 again, normal shifting may begin once more.

There are other ways to load parallel data into a shift register. But this particular method should give you an idea of the kinds of extra features that can be added to the basic shift register. By providing appropriate control signals and gating, you can even make a "bidirectional" shift register that will shift either forward or backward.

What are some other kinds of flip-flops?

R-S flip-flops are limited because their output is unpredictable if both the R and S inputs are a 1 at the same time. This condition must be avoided in any circuit design.

R-S flip-flops are relatively simple, economical units that work just fine in many applications – especially plain shift registers. However, they have a peculiar quirk that's illustrated in the truth table shown in *Figure 4-7*. Note that a truth table for a *clocked* flip-flop like this one shows the input states *before* a clock pulse, and the "true" output *after* the next clock pulse. (For example, the first line in the table says that if S and R are both 0 to begin with, then after the next clock pulse, Q will stay the same as it was before.)

**Figure 4-7.
Symbol and Truth Table for a Clocked R-S Flip-Flop**

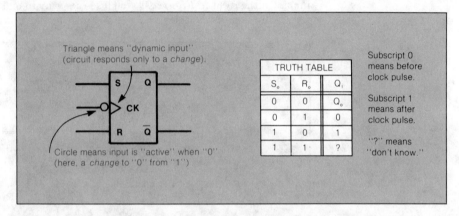

The peculiar quirk we're talking about is that when S and R are both 1, then after a clock pulse, the output could be either 1 or 0. There's *no telling* what it will be. So you *never* let a clock pulse occur while S and R are both 1.

Other kinds of flip-flops don't have this problem. There are three other types commonly used, so let's take a quick look at them.

"D" flip-flops

In a D flip-flop, whatever state is present on the D input before the clock pulse, will appear on the Q output after the clock pulse.

Figure 4-8 shows the truth table for all "D" flip-flops. (D stands for *data* or *delay.*)The true output after a clock pulse is held in the state maintained at the input, D, during the clock pulse. Note that the outputs change *only* at the moment the clock goes from 1 to 0. (Changes at D have *no effect* while the clock is 1, as in the gated latch of *Figure 4-1.*) There are any number of ways to make a D-type flip-flop. On the left in *Figure 4-8*, you can see how to make one out of an R-S master-slave flip-flop and an inverter. Preset and clear inputs could be provided, but are not shown here.

**Figure 4-8.
A D Flip-Flop**

Changes here have no effect while the clock is at 1.

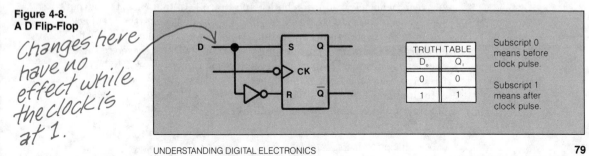

"T" flip-flops

The T flip-flop changes the state of its output at each clock signal. It toggles from one state to the other; thus the name T flip-flop.

A "T" or "toggle" flip-flop is one that has no data inputs at all. Its outputs simply *change state* at every clock pulse (either when the clock goes to 1 or when it goes to 0, depending on the design of the particular flip-flop. We say that the flip-flop "toggles" at every clock pulse. *Figure 4-9* shows how to make a T flip-flop by feeding back the outputs of an R-S master-slave flip-flop to the inputs (Q to R, and $\overline{Q}$ to S).

This particular T flip-flop toggles when the clock input (now labeled "T") goes from 1 to 0, as indicated by the little circle at the clock input. On the right in *Figure 4-9*, you'll see one way to make a truth table for a T flip-flop that toggles on a 1-to-0 transition like this one.

**Figure 4-9.
A T Flip-Flop**

When the clock input goes to 0, the flip-flop toggles to the opposite state.

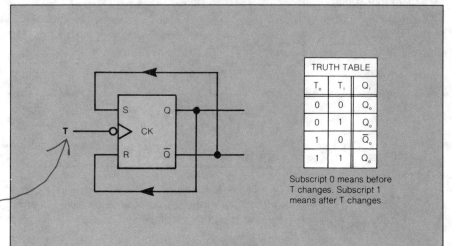

TRUTH TABLE		
T_0	T_1	Q_1
0	0	Q_0
0	1	Q_0
1	0	$\overline{Q}_0$
1	1	Q_0

Subscript 0 means before T changes. Subscript 1 means after T changes.

"J-K" flip-flops

Figure 4-10 shows one way to make a "J-K" flip-flop out of an R-S master-slave flip-flop like the one in *Figure 4-5*. (Preset and clear inputs are not shown.) The feedback pattern here is like that in *Figure 4-9*, except that *extra inputs* are provided to the AND gates in the front end of the flip-flop. The "true" input is called J, and the "complement" input is called K. (Some people say these letters come from the names of two men who developed this idea.)

The J-K flip-flop operates like a R-S flip-flop except when both J and K inputs are in a 1 state, the output toggles. This avoids the indeterminate state of the R-S flip-flop.

As shown in the truth table in *Figure 4-10* for all J-K flip-flops, this building-block performs exactly like a clocked R-S flip-flop *(Figure 4-7)*, except when clocked while both J and K are 1. In that case, the outputs simply *toggle* to the opposite states, like a T flip-flop. Thus, the feedback prevents the troublesome "uncertain" condition from occurring. J-K flip-flops are very versatile, popular units, which we will shortly see in some circuit applications. If a particular application calls for J or K to be the AND function of several signals, these signals can simply be brought in as extra inputs to the AND gates in the front end of the flip-flop. These multiple J and K inputs are called J_1, J_2, K_1, K_2, and so forth.

**Figure 4-10.
A J-K Flip-Flop**

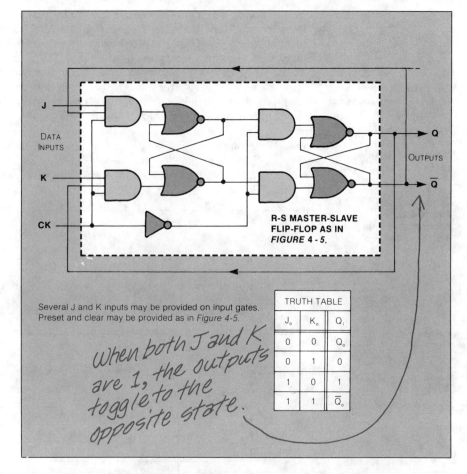

DATA
INPUTS

J

K

CK

OUTPUTS

Q

$\overline{Q}$

R-S MASTER-SLAVE
FLIP-FLOP AS IN
FIGURE 4 - 5.

Several J and K inputs may be provided on input gates.
Preset and clear may be provided as in *Figure 4-5.*

*When both J and K
are 1, the outputs
toggle to the
opposite state.*

TRUTH TABLE		
J_o	K_o	Q_1
0	0	Q_o
0	1	0
1	0	1
1	1	$\overline{Q}_o$

There are a number of variations on the four basic types of flip-flops (R-S, D, T, and J-K). Most of the variations pertain to clocking – that is, to exactly what happens when the clock goes from 0 to 1, when it's at 1, when it goes from 1 to 0, and when it's at 0. These variations don't need to be of any concern to us in this book. You can look into them further by yourself when the need arises.

How do you make an "asynchronous binary counter"?

Now that you're familiar with the various basic kinds of flip-flops and how they're used in parallel and shift registers, let's move on to one more important kind of sequential building block, the "counter." There are many different kinds of counters. But all of them can be considered to be a special kind of register with one input and usually with a parallel output from each flip-flop. The network *counts pulses arriving at the input*, and stores the total count in the flip-flops. If parallel outputs are provided, they show the stored total in a numerical code. The code is usually in the form of binary numbers.

Figure 4-11 shows how to use four T flip-flops to make a "4-bit binary counter" (similar to the Texas Instruments SN7493A). Every time the input signal goes from 1 to 0, the counter registers the next higher binary 4-bit number. The next count after 15 (1111) is zero (0000). The counter can be cleared to zero by means of a common "clear" input.

The binary counter is another important sequential building block. A simple asynchronous binary counter is made by connecting the output of a T flip-flop to the input of another T flip-flop that follows it.

The way the counter works is by having the "true" output Q of each flip-flop (except the last one on the left) connected to the "toggle" input of the next flip-flop to the left. Every time one of these outputs goes from 1 to 0, the next flip-flop to the left toggles to the opposite state, as you can see in the signal diagrams below each output and the input. This action makes the outputs actually *count upwards* in binary code. As you can imagine, a 5-bit counter would count from 0 to 31, a 6-bit counter from 0 to 63, and so on.

Figure 4-11.
An Asynchronous Binary
4-bit Counter

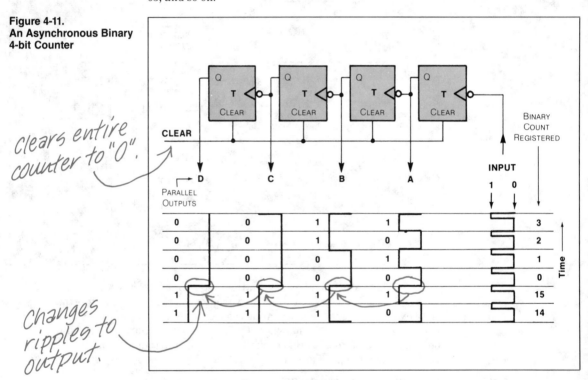

Clears entire counter to "0".

Changes ripples to output.

Why is this counter called "asynchronous"?

The counter in *Figure 4-11* is called an "asynchronous" type. This is because the output changes are not "synchronized," meaning the outputs don't change at *exactly* the same time. For example, when the count changes from 15 to 0 (see the signal diagrams in *Figure 4-11*), it takes a certain small amount of *time* for each flip-flop to go to 0. So the changes "ripple through" the counter from right to left. (For this reason, an asynchronous counter is often called a "ripple counter.") During the time these changes are rippling through, *the output count is not correct.*

"Ripple-thru" asynchronous counter outputs do not change state at exactly the same time but change state in a sequence or wave as the output of one flip-flop triggers the input of the one to which it is coupled, and so on down the line.

One application in which the rippling causes no problems is when the only output used is the last one on the left — output D in the case of a 4-bit counter. This output provides pulses with a frequency only one-sixteenth as fast as the input pulses. "Frequency division" is a useful operation in many digital systems. The counter in *Figure 4-11*, would be called a "divide-by-16" counter, or a "modulo-16" counter. (We'll come back to this word "modulo" in a moment.)

How can you make a "synchronous" counter?

On the other hand, many applications require that the outputs of a counter change at *exactly* the same time (or at least, as nearly exactly as digital circuitry can make them do so). Such a counter is called "synchronous." *Figure 4-12* shows one way to make a synchronous binary 4-bit counter, using four J-K flip-flops. This sequential building-block shows how versatile and useful a J-K flip-flop can be.

**Figure 4-12.
Synchronous Binary 4-bit Counter**

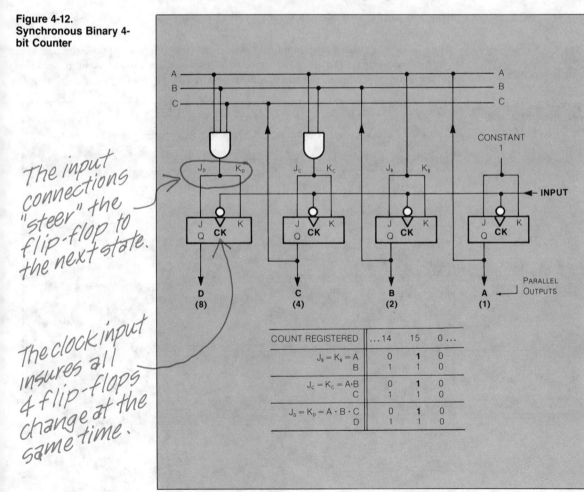

The input connections "steer" the flip-flop to the next state.

The clock input insures all 4 flip-flops change at the same time.

COUNT REGISTERED		...14	15	0...
$J_B = K_B = A$		0	1	0
	B	1	1	0
$J_C = K_C = A \cdot B$		0	1	0
	C	1	1	0
$J_D = K_D = A \cdot B \cdot C$		0	1	0
	D	1	1	0

Synchronous counters use clock pulses as the input and each counter's flip-flop state, after the clock pulse, is determined by its state before the clock pulse and logic circuits called steering circuits.

The input carrying pulses to be counted goes to the *clock input* of all four flip-flops, which guarantees they all change at the same time. The particular state each flip-flop goes to after each clock pulse is determined by the way the J and K inputs are connected. We say that the input connections "steer" each flip-flop to the correct next state. Note that the J and K inputs of each flip-flop are tied together. As we saw in the J-K truth table in *Figure 4-10*, when both inputs are 0, the flip-flop remains in its present state. But when both inputs are 1, it *toggles* at the next clock pulse. (You may want to trace the operation through each count, referring to the J-K truth table. A table like the one shown partially in *Figure 4-12* will be helpful.)

Notice the similarity between the steering circuitry and that of the *decoder* -type networks we've seen in Chapter 3. Steering a flip-flop to the correct next state is simply a matter of *decoding* (recognizing) the state immediately before you want the flip-flop to change.

How can you make a counter count to other numbers?

Not all counters count only 16 states — or 4, or 8, or 32, or some other "power of two." For example, *Figure 4-13* shows one way to make a synchronous counter that counts from zero to nine and back to zero, using binary code. It's called a "decade" counter, or a "BCD" counter.

**Figure 4-13.
Synchronous Decade
(BCD) Counter**

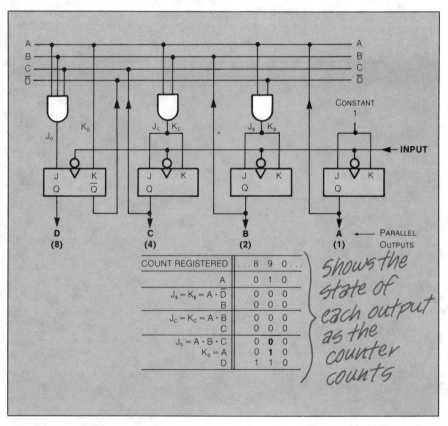

COUNT REGISTERED		. . 8	9	0 . . .
	A	0	1	0
$J_B = K_B = A \cdot \overline{D}$		0	0	0
	B	0	0	0
$J_C = K_C = A \cdot B$		0	0	0
	C	0	0	0
$J_D = A \cdot B \cdot C$		0	**0**	0
$K_D = A$		0	**1**	0
	D	1	1	0

Shows the state of each output as the counter counts

Using different steering circuits, counters can be designed to count to many different modulus (the number of states it counts through).

As you can see, this counter looks much like the synchronous 4-bit binary counter in the preceding figure. However, the "steering" circuitry is different. It sends the counter back to zero on the count after "nine." We say that the counter's "modulus" has been shortened from 16 to 10. The modulus of a counter is the number of different states it counts through. A counter with ten states like this one in *Figure 4-13* is called a "modulo-10" counter. By using enough flip-flops and designing the proper steering circuitry, you can make a counter with any modulus you need for a particular application. (Once again, *Figure 4-13* includes part of a table you may want to make out to help you trace the operation of this counter through each count.)

How does a digital watch or clock use counters?

Perhaps the most familiar application of counters to most people is in a digital watch or clock. *Figure 4-14* shows how counters could be used in a watch or portable clock system that displays hours, minutes, and seconds in a 12-hour cycle.

**Figure 4-14.
Counters in a Digital Watch**

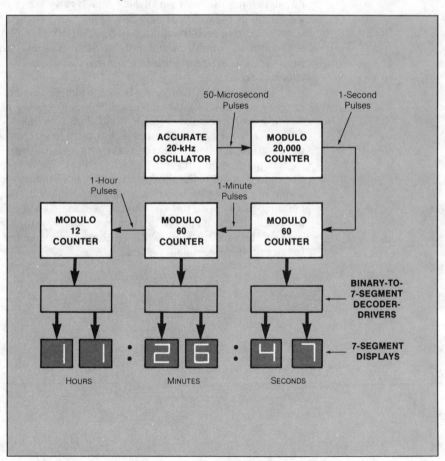

An oscillator circuit using a special vibrating crystal produces sharply square pulses at a very accurate frequency of 20 kilohertz (20,000 cycles per second). A counter with a modulus of 20,000 counts these pulses and puts out a pulse every second. A series of three more counters counts seconds, minutes, and hours by operating with the appropriate modulus in each counter. The parallel binary output from each of these counters is decoded in order to drive a 7-segment display for each digit of the hours, minutes, and seconds. Incidentally, none of these counters needs to be synchronous, since the rippling of states in a counter would occur much faster than your eye could see in the display. And the accuracy of frequency division is not affected by rippling.

Are there any other digital building-blocks?

As there are many different kinds of registers, there are also any number of variations on the basic counter principles we've seen. For example, some counters can be preset to a particular state, like parallel data being entered into a shift register. And counters sometimes have decoders built into their parallel outputs. But we've seen enough to give us a general idea of what counters do and how they work.

And in fact, the same goes for *all* digital building-blocks. The sequential and combinational units we've seen in this chapter and the preceding one are representative of nearly all digital building-blocks you'll ever run across.

We've come a long way from the simplest building-blocks – the gates. We've seen how the combinational requirements of a particular building-block can be analyzed to decide how to connect a network of gates. We've seen how gates can be arranged in such a way as to *store* information, and how we have to take into account the *sequence* of events in time in using building-blocks with memory. We've become familiar with the main types of building-blocks that are used to make a digital system. Along the way, we've looked at a number of applications of various building-blocks in digital systems. We will build upon some of this understanding in a later chapter on mass memory units. Still later, we will see how the building-blocks with which we've become familiar are put together to form complete systems.

Quiz for Chapter 4

1. What is the purpose of clock
   signals in sequential
   building-blocks?
   a. To tell how much time has
   elapsed since the system was
   turned on.
   b. To synchronize the changing
   of data in various parts of a
   system.
   c. To carry serial data signals.
   d. All of the above.

2. What keeps the outputs of an
   R-S latch from changing when
   one input changes to 0 while the
   other remains at 0?
   a. The outputs are fed back to
   these two input wires, keeping
   them from changing.
   b. The NOR gates are special
   sequential gates that store
   information inside them.
   c. The output from each NOR
   gate is fed back to another
   input on the other gate.
   d. B and C above.

3. When output Q from an R-S
   latch is 0, and both inputs R and
   S are 0, then a momentary "1"
   pulse at S will:
   a. Flip Q from 0 to 1.
   b. Flip Q from 0 to 1 and back
   to 0.
   c. Flip $\overline{Q}$ from 0 to 1.
   d. Flip $\overline{Q}$ from 0 to 1 and back
   to 0.

4. A group of bits transmitted and
   received at the same time is
   called:
   a. A clock signal.
   b. Parallel data.
   c. Serial data.
   d. B and C above.

5. Which kind of register can
   convert data from serial to
   parallel form, or parallel to
   serial?
   a. Parallel register.
   b. Shift register.
   c. Synchronous counter.
   d. Asynchronous counter.

6. A parallel register is used for:
   a. Temporarily holding parallel
   data.
   b. Temporarily holding serial
   data.
   c. Shifting parallel data.
   d. Shifting serial data.

7. Do all flip-flops in a register
   receive the same clock signal?
   a. Yes.
   b. No.
   c. Only in the case of parallel
   registers.
   d. Only in the case of shift
   registers.

8. What is the purpose of the
   "master-slave" feature in a
   flip-flop?
   a. To let it change its output
   before accepting a new input.
   b. To make it simpler and more
   economical.
   c. To let it accept a new data
   input before changing its
   output.
   d. B and C above.

9. A momentary signal at the "preset" input of a flip-flop will:
   a. Preset the true output to 0.
   b. Preset the true output to 1.
   c. Clock in new data from the data inputs.
   d. Any of the above.

10. The peculiar quirk that R-S flip-flops have is that when one is clocked while both R and S are 1:
    a. The outputs don't change.
    b. The outputs toggle to the opposite state.
    c. The true output is cleared to 0.
    d. You can't tell what the outputs will be.

11. Which kind of flip-flop has only one data input?
    a. "R-S"
    b. "D"
    c. "T"
    d. "J-K"

12. Which kind of flip-flop changes its outputs to the opposite state at every input pulse?
    a. "R-S"
    b. "D"
    c. "T"
    d. "J-K"

13. A J-K flip-flop behaves like an R-S flip-flop except that when clocked while J and K are both 1:
    a. The outputs are uncertain.
    b. The outputs do not change.
    c. The outputs both go to "1."
    d. The outputs toggle to the opposite states.

14. An asynchronous binary counter:
    a. Is made of T flip-flops.
    b. Changes its outputs in "ripple" fashion.
    c. Is useful as a frequency divider.
    d. All of the above.

15. A counter can be made synchronous, or its modulus can be shortened, or both, by:
    a. Providing decoded outputs.
    b. "Steering" the clock inputs to the desired states.
    c. "Steering" each flip-flop to the correct next state by decoding the state before.
    d. All of the above.

(Answers in back of the book)

Why Digital?

Now that you have in your mind the main outlines of the comprehensive picture we're painting of digital electronics, this is a good time for us to pause for a while and evaluate the nature of the digital approach to building a system. In speaking of the *only other* general way to build a system (in Chapter 1), we used the words "analog" and "linear." So first, we have to explore these ideas and their meanings. This new understanding will enable us to see more clearly *why* the digital approach is used in certain situations—and perhaps more important, why it's *not* used in other situations.

What is analog information?

All we have noted so far about the analog (or linear) method was with regard to our discussion of the universal system organization back in *Figure 1-20. Figure 5-1* shows this organization again as it applies to analog systems. We've shown that *digital* systems manipulate *information in digital form.* That is, the information is made up of separate parts, or bits. We added that another way to represent information is by a method called *analog.*

This method of handling (manipulating or transmitting) information is what makes analog systems different from digital systems. "Linear" is a name sometimes used for the general type of electronic *circuitry* used to *handle* analog information. So let's see what's meant by analog information.

The main difference between digital and analog systems is in the signal that is used to transmit and manipulate information.

**Figure 5-1.
Analog Systems Use
"Linear" Circuits**

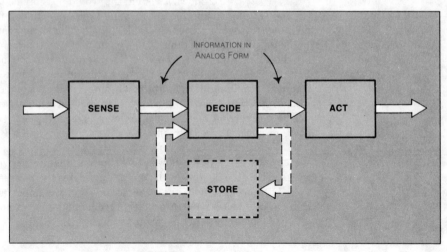

In an analog system, an electrical signal is used to represent or model a physical quantity.

To put it briefly, in an electrical analog system, we use some *controllable property of electricity*, such as current or voltage, as an "analog" to represent the information we're handling. (Think of the word "analogy.") That is, the electricity is closely and carefully controlled so as to be a more or less exact copy or representation or model of the information. *Figure 5-2* shows an example.

**Figure 5-2.
A Fuel-Level Analog
System**

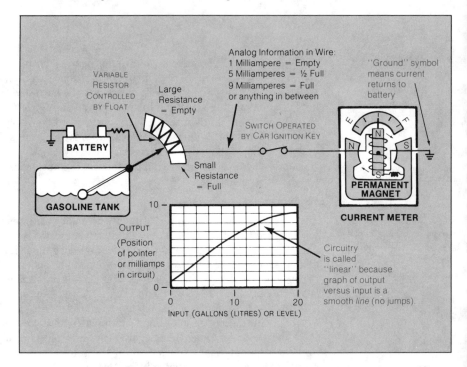

How can an electric fuel gage use analog information?

Figure 5-2 shows a system that could possibly be used to indicate the fuel level in an automobile gasoline tank in an analog fashion. The float on a swinging arm in the tank adjusts a "variable resistor" according to the gasoline level. (Think of a lever turning the volume-control knob on a radio.) This *varies the current* in a wire running to the instrument panel. For example, as shown in the figure, 1 milliampere of current might mean "empty," 9 milliamperes might mean "full," and each current in between would represent a certain tank level. At the instrument panel, the pointer needle in a current meter indicates level as it moves between the two extremes of measured current.

Note that the current is *not switched on and off* as in a digital system. Instead, current flows at all times and is *varied* over a certain range. Thus, the pointer in the meter duplicates the movement of the float over its full range, more or less accurately. What's important to notice is that *current* has carried information from one place to another in an *analog* fashion, by being *varied* over a range.

In what way is this analog circuit "linear"?

This "smoothly varying" feature of analog systems is illustrated in the graph in *Figure 5-2*, which also shows us where we get the name "linear" for circuits that handle analog information. This graph shows generally how the circuit *output* (pointer position or milliamps) changes when the *input* (gallons–litres–or level) changes. Since the current is *varied* rather than being switched, the graph is a *smooth line*, with no sudden jumps. That's why we call the circuit "linear."

Unfortunately, the word "linear" can have different meanings in different circumstances. Sometimes it's used to mean that a graph is not only smooth but perfectly *straight*. Although you need to be acquainted with this word, we will not be using it very much in this book, but will mainly refer to "analog" circuitry.

How do the parts of this system work?

In case you're not familiar with the parts of the system in *Figure 5-2*, we ought to say a few words about how they work. The variable resistor would typically be a curved piece of carbon touched by a moveable contact. Sliding this contact along the carbon shortens or lengthens the path travelled by the electricity through the carbon. Another resistor, shown further left, holds the current down to the range the meter can handle. The "ground" symbols at the battery and meter indicate that these two points are connected through a "common" path, such as the frame of the car, which is considered to be at zero volts. (As you probably already know, a *complete circuit* must be provided before current will flow.) A switch turns off the current when the ignition key is off.

The current meter acts as a little electric motor whose rotor is kept from turning very far by a spring. The greater the current, the stronger the magnetic field produced in the little rotating coil, and the more strongly the north and south poles of the coil and the permanent magnet interact with each other. (Remember, "unlike" poles attract and "like" poles repel each other.) As the coil turns against the force of the spring, it moves the pointer to indicate current through the coil.

What has the fuel gage illustrated?

This, then, is an example of how one controllable property of electricity–namely, current–can be used as an "analog" (a direct representation) of the information we want to transmit. It's a very simple, idea, actually. It's the natural way anyone would think of to handle information by means of electricity.

This example may have led you to realize that many of the common everyday electrical systems you can think of use some form of analog information, either throughout the system or in part of it. To learn more, let's look at a couple more analog systems you may be familiar with.

Linear does not necessarily mean the signal varies in a straight line relationship. Rather, the signal varies in a smooth continuous wave-like form not in on off levels like a digital signal.

In the fuel gauge, the current in the circuit is an analog of the fluid level in the tank. If the fluid level goes down the current reduces. When the current reduces the meter reads closer to zero.

In a telephone circuit, the circuit current is an analog of the sound hitting the microphone. When the current flows through the coil of the earpiece it causes the diaphragm to move and reproduce the sound.

How does a telephone use analog information?

Although a telephone system may seem pretty far removed from the current-analog fuel-gage system, it works essentially the same way, only much *faster*. *Figure 5-3* shows the idea of a simplified telephone system, consisting of one mouthpiece and one earpiece. Current flows through the complete circuit, from the direct-current power supply, through the mouthpiece and earpiece, and back to the power supply through the ground connections. As in the fuel-gage system, the amount of current is determined by the resistance in the circuit.

The microphone element in the mouthpiece is a capsule full of powdered carbon that acts as a *variable resistor*. It allows more current to pass when it is squeezed by air pressure. When we speak, we vary the air pressure in front of our mouths. Fluctuations in air pressure occurring from about 20 times a second (a "frequency" of 20 "hertz") to about 10,000 times a second (a frequency of 10 "kilohertz") are what our ears hear as *sound*.

So the microphone element creates rapidly surging waves of electricity in its output wire, as an analog representation of waves of varying air pressure—that is, sound waves. (The current increases and decreases very quickly, over and over again, due to the varying resistance.) At the earpiece, this current passes through the coil of a fixed electromagnet, creating surges of magnetic force that match the surges of current. The magnetic force, in turn, attracts a springy metal diaphragm in proportion to the current waves. Finally, the diaphragm, by rapidly pushing and pulling the air, reconstructs a more or less accurate copy of the original sound waves, which have been transmitted in analog form by surges of electricity. Our ears detect the varying air pressure and hear the sound. Simple, isn't it?

**Figure 5-3.
An Analog Telephone
System**

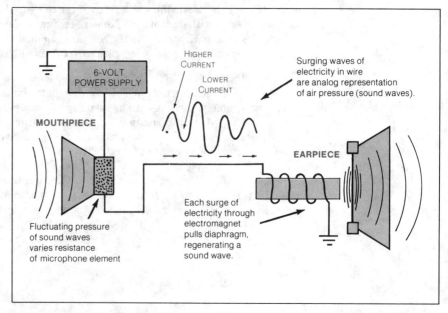

What's the difference between current and voltage analog?

In most of the other examples of analog systems that we'll study, we'll be speaking of controlling the *voltage* of electricity rather than the current. Thus, we will be using *voltage analog* rather than current analog. These two methods are so similar in concept that the difference need not concern us here. Current is simply the amount or volume of electricity flowing, while voltage is the pressure of electricity—the force that moves the current.

How do radio systems use "amplitude modulation"?

In amplitude modulation (AM) radio systems, the sound waves change the amplitude of a radio wave in a pattern matching the sound waves.

Although current and voltage are the basic workhorses in the analog world, these methods are often used in some very special ways known by different names—ways that make the analog method even more useful. Let's look at a familiar example. This will be an AM radio transmitter such as those used in citizens-band radios, "ham" radios, and regular radio broadcasts. "AM" stands for "amplitude modulation," which is one of the more advanced forms of analog information.

Figure 5-4 shows how this technique lets us transmit *sound* waves (whose frequencies are relatively low, from 20 to 10,000 waves per second) by using *radio* waves of a much higher frequency (say, 1 megahertz—a million waves per second). The transmitter simply "modulates" (that is, varies) the "amplitude" (the height or strength) of the radio waves in a pattern matching sound waves. That's what amplitude modulation means. (To "modulate" something means to vary or change it to fit a certain pattern. We could call current analog "current modulation.")

**Figure 5-4.
An AM Radio Transmitter**

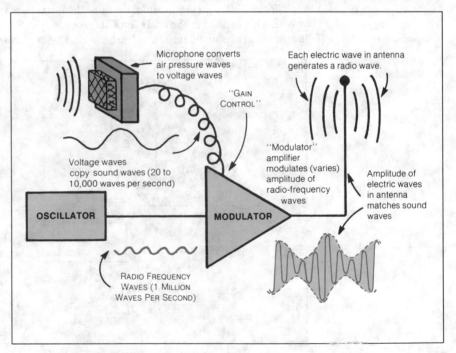

Microphone converts air pressure waves to voltage waves

"GAIN CONTROL"

Each electric wave in antenna generates a radio wave.

Voltage waves copy sound waves (20 to 10,000 waves per second)

"Modulator" amplifier modulates (varies) amplitude of radio-frequency waves

Amplitude of electric waves in antenna matches sound waves

OSCILLATOR

MODULATOR

RADIO FREQUENCY WAVES (1 MILLION WAVES PER SECOND)

To perform this trick, as shown in *Figure 5-4*, the system uses an "oscillator" to generate 1-megahertz electrical waves of *constant* amplitude. (The voltage is just going up and down smoothly a million times a second, the *same amount* each time.) A special "amplifier" circuit called a "modulator" then amplifies these waves. That is, it *multiplies* its input voltage by a certain factor, producing taller and stronger waves at its output. (The triangle you see is the customary symbol used for an amplifier when you're not concerned with what's inside.)

The Amplitude Modulation (AM) is accomplished by changing the gain of a high frequency carrier amplifier according to the voltage analog of the sound waves produced by the microphone.

Now the "gain" of the amplifier—the factor it multiplies the input by—is controlled by the voltage signal from a microphone. This signal is a voltage analog of the *sound waves* striking the microphone. The voltage waves, in effect, rapidly "turn the volume control knob" on the amplifier up and down, thus modulating the amplitude of the 1-megahertz output waves as we desire.

On reaching the antenna, each electric wave generates a radio wave. The radio waves, in turn, are of the same frequency and amplitude (relative to one another) as the amplitude-modulated electric waves. Thus, by modulating the amplitude of radio waves, we can transmit sound in an analog fashion.

We won't go into how a radio *receiver* works. It's enough for us to say that it responds only to radio waves of the frequency it is tuned to, and that it recovers the original sound-wave pattern by following just the *peaks* of the 1-megahertz waves.

How does "FM" radio transmit information?

In frequency modulation (FM), the microphone signals change the frequency of the carrier signal according to the sound rather than the amplitude of the carrier as in AM.

Another variation of the analog technique, similar to amplitude modulation, is *frequency modulation* of both electrical waves and radio waves. This method is the basis for FM (frequency-modulation) radio communications, including broadcast FM and television sound.

**Figure 5-5.
FM Modulation**

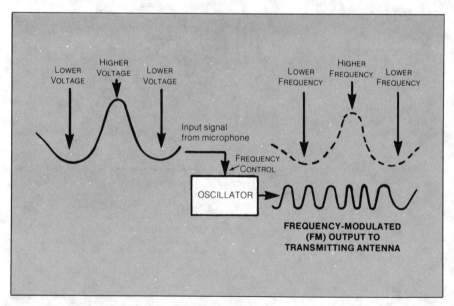

As indicated in *Figure 5-5*, frequency-modulated electric waves are produced by *controlling the frequency* at which an oscillator varies its output voltage. If these waves are sent to an antenna, they produce corresponding FM radio waves. FM signals are less subject to interference by noise, because noise mainly affects the signal *amplitude* and thus does not bother the *frequency* signals. However, FM waves are more troublesome to generate and receive.

**Figure 5-6.
Analog Systems
Decisions/Storage**

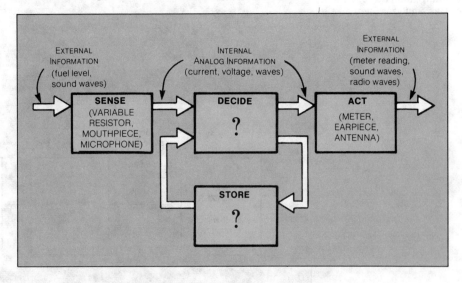

Which "universal system functions" have been illustrated?

Sensors in analog systems convert physical quantities to electrical input signals to perform the sense function. Actuators convert the electrical signals at the output to action.

As a reminder of what we've learned so far about analog systems, and to see where to go from here, let's consider how the systems we've studied fit the universal system organization, as shown in *Figure 5-6*.

The variable resistor, mouthpiece, and microphone *sense* external information and convert it into analog form by varying current or voltage. And the meter, earpiece, and antenna *act* to convert varying electricity into meter indications, sound waves, and radio waves. So we pretty well understand these two stages of analog systems.

But what about the "decide" and "storage" functions in analog systems?

How do analog circuits usually store information?

Let's consider storage first. Analog signals can be stored for a very short, fixed length of time by *delaying* the signal. This method involves sending the signal on a "detour" through a special path in which the signal travels much more *slowly* than it would through a wire. (In a wire, changes in signal level travel at nearly the speed of light.) Thus, the analog information is stored for the period of time the information is travelling in the slow path. These slow paths are called "delay lines." A delay line can store only a small amount of information at a time, and for only a fraction of a second.

To store analog information for a period as long as several minutes, a system usually stores electric *charge* at a certain *voltage level*, using a device called a capacitor. For an example of how this works, imagine an automatic gas-analyzer system in a chemical plant, shown in *Figure 5-7*.

**Figure 5-7.
Information Storage for a
Gas Analyzer System**

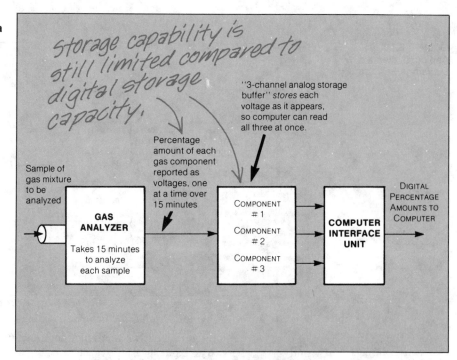

Storage capability is still limited compared to digital storage capacity.

"3-channel analog storage buffer" *stores* each voltage as it appears, so computer can read all three at once.

Percentage amount of each gas component reported as voltages, one at a time over 15 minutes

Sample of gas mixture to be analyzed

GAS ANALYZER

Takes 15 minutes to analyze each sample

COMPONENT #1

COMPONENT #2

COMPONENT #3

COMPUTER INTERFACE UNIT

DIGITAL PERCENTAGE AMOUNTS TO COMPUTER

How does a capacitor store analog information?

The gas analyzer analyzes a sample of a gas mixture every 15 minutes, and reports the amounts of three different gases that are components (parts) of the mixture. The percentage amount of each component appears as a voltage-analog signal at the analyzer output for just a few seconds. The three voltage signals are produced one at a time over an interval of 15 minutes for each complete analysis. We need a "three-channel analog storage buffer" unit to *store* these three voltage signals from each analysis. This is so that after the final voltage appears, all three can be transmitted (sent) quickly to a computer for recording, through a computer-interface unit.

Figure 5-8 shows one of the three voltage-storing channels. As soon as a new voltage is ready for storage, the analyzer energizes the electromagnet in a "relay" for a short time. The magnetic field attracts a metal arm and closes the relay switch contacts, connecting the voltage output to the capacitor for a few seconds. As a result, current flows between the analyzer output and the capacitor, causing electric charge to be stored by the capacitor to the voltage input.

By charging a capacitor to the input voltage of a system at particular times (sampling times), the input information can be stored for later use by the analog system.

As indicated in *Figure 5-8*, capacitors consist essentially of two closely-spaced metal plates with a thin layer of insulation to prevent current from flowing between them. They serve many different functions in electrical circuits, but this one acts as a *storage reservoir for electric charge*. (Imagine a tank storing compressed air at a certain pressure.) The more charge we pack into the capacitor, as indicated by the little "plus" signs, the higher the voltage that we store. After the capacitor is charged up to the voltage level of the input wire, the relay contacts are opened, disconnecting the capacitor. So we actually *store the input voltage* until the relay is turned on again to store a new input voltage.

Figure 5-8.
Charging a Capacitor to Store a Voltage Level

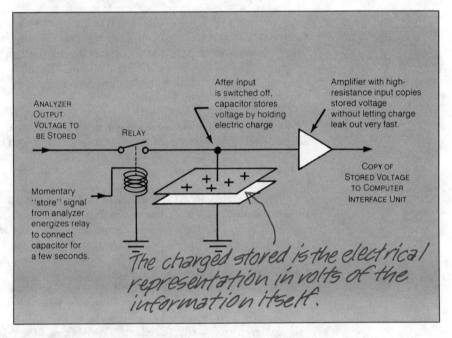

The important job of "reading" the stored voltage without changing it too much, and sending a close copy to the computer-interface unit, is handled by a special amplifier circuit. (Like all amplifiers, it's shown as a triangle.) This amplifier has a very high resistance in its input. Therefore, it causes only a very small current to be drained from the capacitor as the amplifier "senses" and copies the capacitor voltage.

An amplifier with very high resistance at its input senses the stored voltage, amplifies it and passes it on to the system. If any charge is drained off the capacitor, it causes an error.

Unfortunately, even this small current will *drain charge* from the capacitor and lower the voltage. Over the storage period of 15 minutes, the amplifier output voltage will *decay* gradually from the correct value—perhaps as much as five percent. The decay of voltage stored in a capacitor can be a serious source of error in some applications where accuracy is important. But as we said earlier, this is just about the only practical way to store analog information any longer than a fraction of a second. Obviously, storage is one of the big problems you run into in using analog methods.

How do analog circuits make decisions?

As for the "decide" function in analog systems, we can take care of this matter rather quickly. The fact is that we have *already* seen analog decisions being made in our examples, without taking note of it. Let's look back at some of the example figures.

In *Figure 5-2*, the float and variable resistor in the fuel-gage determine what current to transmit in response to a certain fuel level. This process can be considered a sort of *decision*. The meter, in turn, decides what level to indicate for a certain current. Similar decisions are made by the microphone element and the earpiece in the telephone system in *Figure 5-3*.

In the AM radio transmitter in *Figure 5-4*, the modulator-amplifier actually performs a *voltage multiplication* decision. At each and every moment of time, it multiplies the input voltage by a factor controlled by the "gain" signal provided by the microphone. When you stop and think about it, multiplication is a pretty respectable decision for anyone or anything to make!

Of course, you can see what we're driving at. Whenever electricity is modified in some fashion in an analog circuit, *information is being processed*. Existing information is being employed to create new information, or new forms of information. This is the action we have been calling "deciding," to emphasize its importance with regard to information.

It's important for you to note that this analog deciding process is not made up of separate steps as in a digital system, but is a *continuous process*. When electronic devices such as transistors are involved (as in amplifiers), the devices do not switch on and off. Instead, they *vary* the flow of current *in between* the "on" and "off" states, in a smooth fashion we called "linear" early in this chapter.

When a transistor is operated in this "in-between" range, it acts as an electrically-controlled *variable resistor*. When used this way in an electric circuit, a transistor is an *amplifying* element. A *small* change in the control signal varies the effective resistance of the transistor, producing a larger or "amplified" change in output current. Transistor amplifiers of various kinds are the main building-blocks in electronic analog systems.

What are the advantages of digital methods?

Finally, we've covered enough details of analog methods to begin making some direct comparisons between digital and analog, so we can see why digital techniques are used in certain situations. First, we'll consider the relative *advantages* of digital methods, then later come around to the *limitations*. For your reference, the points we'll cover are listed in *Figure 5-9*.

Analog systems are performing the "decide" function when they convert and modify the electrical form of signals as they handle information. However, the signals are always varying in a continuous manner rather than in discrete steps as in digital.

**Figure 5-9.
Comparing Digital and
Analog Methods**

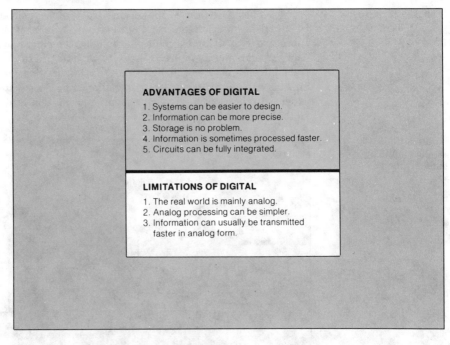

ADVANTAGES OF DIGITAL

1. Systems can be easier to design.
2. Information can be more precise.
3. Storage is no problem.
4. Information is sometimes processed faster.
5. Circuits can be fully integrated.

LIMITATIONS OF DIGITAL

1. The real world is mainly analog.
2. Analog processing can be simpler.
3. Information can usually be transmitted
   faster in analog form.

1. Digital systems can be easier to design

Binary digital systems are designed using only two electrical levels or states. As a result, the circuits are much simpler and have less rigid specifications than analog circuits.

As we have already noted, in analyzing and designing *digital* systems, our only direct concern with electricity is whether it's "on" or "off." We don't have to worry about *exactly* what voltage or current is in a wire. All we care about is that it's not "in between" the two permitted states. Consequently, the circuits we work with—*switching* circuits—can be much *simpler* than analog circuits, and the devices in the circuits don't have to fit such close specifications.

Furthermore, as you have begun to see, digital systems are all built up out of a small handful of basic building-block circuits—gates and flip-flops—and larger building-blocks made from them (decoders, counters, etc.) Within a given system or subsystem, all the gates and flip-flops are usually members of the same "family" of digital circuits, such as TTL, MOS, and others. As we will see in the next chapter, this means the circuits resemble one another closely. Consequently, the building-blocks are all perfectly *compatible* with one another, provided the designer observes a few simple rules. He can, in effect, put together a digital system like assembling tinkertoys.

2. Digital information can be more precise

Figuratively speaking, every analog system has to grab hold of electricity, wrestle and grapple with it, and bend and twist it to make it match the information that must be transmitted. The result is never a *perfect* analog copy. There's always some *error*, which is expensive and troublesome to reduce.

Digital systems handle numbers much more precisely than analog systems.

Such inaccuracies are permissible in some applications but would be out of the question in others. For example, when you multiply two times two with an analog multiplier (analog computers do it all the time, using an amplifier as we have discussed with regard to the AM transmitter), you aren't likely to get *exactly* four. (See *Figure 5-10*.) You may instead get 3.976, or 4.028, depending on how *accurate* (and how expensive) the amplifier is. Consequently, people don't use analog methods for handling extremely *precise* information.

Figure 5-10.
Digital vs Analog
Multiplication

Analog signals cannot easily represent quantities with high accuracy.

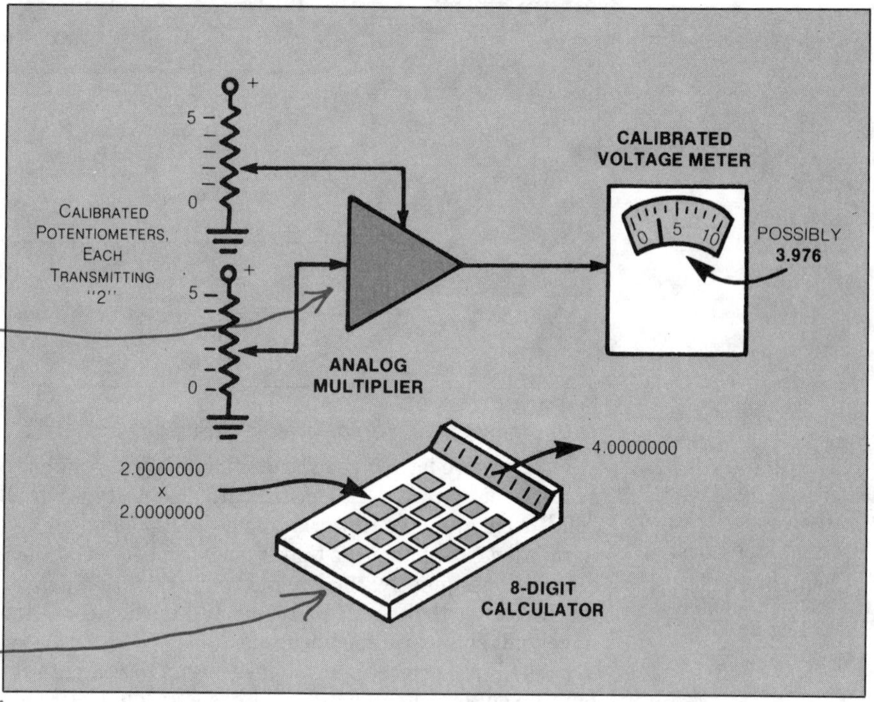

Digital calculations are precise, descrete manipulations

On the other hand, digital methods can handle numbers as long and as precise as you need. Our example hand-held calculator from Chapter 1 handles decimal numbers with eight digits, so we can multiply 2.0000000 times 2.0000000 and get 4.0000000 without any trouble at all. Big computers routinely handle decimal numbers that are much longer, and consequently can be carried to more "decimal places" of precision. Such precision can be handled with much less cost than with analog methods because the same simple digital circuits are used—just more of them, for more bits.

3. Digital storage is no problem

Digital systems can accurately store as much information as needed by simply providing the appropriate number of storage circuits.

The capacitor-storage method we have studied (*Figure 5-8*) is just about the *best practical way* to store electrical *analog* information. You've seen that it's not a perfectly accurate way—because there's no way to prevent a slight trickle of charge from escaping the capacitor.

On the other hand, as we have seen in Chapter 4, we can make a switching circuit "latch onto" a piece of *digital* information and hold it with perfect accuracy for as long as we need. And we can store numbers as long and as precise as we like by simply using as many storage circuits as we need.

If an analog system needs long-term, accurate storage, it has to convert the analog information into *digital* form, and use digital storage techniques.

4. Sometimes digital methods are faster

Digital systems process information many, many times faster than analog systems.

When we consider the *speed* with which circuits handle information, we sometimes run into problems with analog methods. Once again, the trouble is that analog circuits have to manhandle the electricity and whip it into shape. This can take *time* to do, especially when for some reason we have to use large capacitors (as in analog storage) or another class of electric components called "inductors." An inductor is any device that makes the electricity interact with a magnetic field. Any device with a *coil* of wire is an inductor, such as the fuel meter in *Figure 5-2* or the telephone earpiece in *Figure 5-3*.

For example, it may take the better part of a *second of time* to flow enough charge into an analog storage capacitor (as in *Figure 5-8*) to make its voltage close enough to that of the voltage source. By comparison, we can easily make a flip-flop that will store an input signal in a few *nanoseconds* (billionths of a second).

5. Digital circuits can be fully integrated

The significant advantage for digital circuits is that they can use the same type device over and over again to perform many different functions. Therefore, making them in integrated form is much less costly than analog circuits.

By far the most important advantage of digital methods is that digital information-processing circuitry can be *entirely fabricated in integrated-circuit chips,* such as the impressively complex calculator chip we saw back in *Figure 1-3*.

The first four advantages we listed (simple design, accuracy, storage, sometimes speed) were in effect for many years before integrated circuits came along. Those advantages propelled digital techniques into applications in digital computers and a few nooks and crannies in predominantly analog systems (such as for storage, as we have mentioned, and for *switching* analog signals as in telephone dialing and routing).

But when integrated circuits came along, the tremendous advantages they brought (which we will analyze in the next chapter) were applied mainly to *digital* circuits. And as integrated circuits have been progressively improved, they have carried digital methods into a much wider variety of applications than before.

For analog systems, many of the devices required to produce analog signals cannot be made in integrated circuit form.

The reason analog circuitry has not been integrated nearly so fully as digital circuitry goes back again to the fact that analog circuits have to force electricity to match outside information. To do this, analog circuits typically need to use several kinds of *devices that just can't be made out of the silicon of an IC chip*—at least, not very economically. Such devices include inductors (coils and transformers), high-capacitance capacitors, and high-precision resistors. A great deal of progress has been made in designing analog circuitry that *doesn't need* such "non-integrable" devices. But the integration has not been on the grand scale that's much more economically feasible for digital circuits.

To keep the picture in balance, we should point out some important types of "linear" circuits that *have* been very successful in integrated form. There are only very few analog systems or subsystems that don't contain a number of linear integrated circuits.

Operational amplifiers, video amplifiers, voltage regulators, audio amplifiers are linear (analog) circuits that are made successfully and economically as integrated circuits.

By far the most common type of linear IC is the "operational amplifier," or "op-amp." An op-amp is a general-purpose building-block to which you can add a few resistors and capacitors, to make nearly any kind of amplifier you want—as long as the frequencies are below about 1 megahertz. And if you want output signals greater than about 10 volts and 0.1 amp, you can add discrete transistors to the output section. Op-amps are "differential" amplifiers—meaning that they amplify the *difference* between the voltages at two different inputs. The output voltage is around 100,000 times this difference, which for practical purposes is assumed to be an *infinite amount*. This "gain factor" is reduced to the desired value by *feeding back* part of the output signal to the "inverting" (subtracting) input.

Another general-purpose building-block type of linear IC is "video amplifiers," or "wide-band" amplifiers. They can be used at frequencies up to around 100 megahertz—although not down to *zero* hertz (direct current) like op-amps. Their gain is controlled by a voltage signal rather than by the feedback arrangement.

Although power dissipation is a definite limitation in integrated circuits, there are some integrated "power" amplifiers that can put out as much as 5 watts of power. This is enough to drive a small loudspeaker. You can make a respectable little phonograph amplifier with just one of these ICs, plus six or eight discrete resistors and capacitors.

In summary, linear IC amplifiers are very widely used as basic building-blocks or "cores" of various specialized linear circuits. They provide the bulk of the circuitry, and the designer only has to add a few external components to make the circuit perform as he wishes.

What are the limitations of digital methods?

From considering the big advantages of digital methods over analog, you may wonder why they haven't taken over the entire field of electronics. But the fact is that digital techniques have some *inherent limitations* that keep such methods out of certain applications. Let's consider what some of these disadvantages are. They're listed along with the advantages, back in *Figure 5-9*.

1. The real world is mainly analog

Because most real world systems are analog, conversion from analog to digital and back again is necessary when operating with digital systems.

First and possibly most important, the information that goes into and out of most systems is *analog* in nature (or "linear," if you prefer). Stop for a moment and think about the information we're talking about. From examples in this chapter, there are fuel levels, meter readings, sound waves, and radio waves. All of this is *analog* information, in that it *varies* anywhere within a range rather than being limited to definite states like digital information.

The same applies to almost any kind of "natural" information you can think of—temperatures, pressures, weights, intensities, positions, speeds, time, and so forth. You may be accustomed to *expressing* such information in digital form. For example, you may say that you weigh 112 pounds, or maybe 165.3799 pounds if you wanted to be more accurate. But in doing so, you're only giving a *digital approximation* for an *inherently analog* quantity.

If a digital system is to deal with "real-world" information, taking in and putting out analog information, it has to *convert* the input information to digital form before working on it, and then *convert* the digital results back to analog again. Many digital systems do just that. For example (*Figure 5-11*), a "computerized" autopilot on an airplane *takes in analog information* on compass heading and how the airplane is tilted, and *puts out analog information* controlling the rudder and ailerons and elevators to keep the plane flying straight and level. All these inputs have to be converted to and from digital form.

However, converting information between analog and digital forms can be *cumbersome and expensive.* Furthermore, the conversion process always introduces *inaccuracies* and takes a certain amount of *time.* (Time can be a critical factor in some systems.) Furthermore, it may allow too much random, unwanted information we call *"noise"* to leak into the system. In the case of the autopilot, the advantages of digital processing are so desirable that we're *willing to pay the price* for the analog conversions. But in a moment, we'll look at another system where it's obviously better to stick with *analog* processing.

On the other hand, digital processing is a shoo-in for situations where both the inputs and the outputs are *digital* information. The prime example, of course, is systems that handle *numbers* (which are digital of their very essence), such as calculators and computers.

However, the same would go for systems handling *letters* or any other sort of symbols, because such things are also inherently digital. After all, our alphabet can be considered a sort of *number system*—one with 26 numerals rather than ten or two! Likewise, digital outputs are involved in any system that *sequences* events in time, such as the controller for a washing machine. (They used to have a motor-driven analog "clock" that flipped switches in sequence. But the newer ones are a hundred percent digital electronic.)

**Figure 5-11.
Digital Systems Must
Accept Analog Signals**

The analog motions must be translated into digital signals.

Real world motions are analog rather than digital.

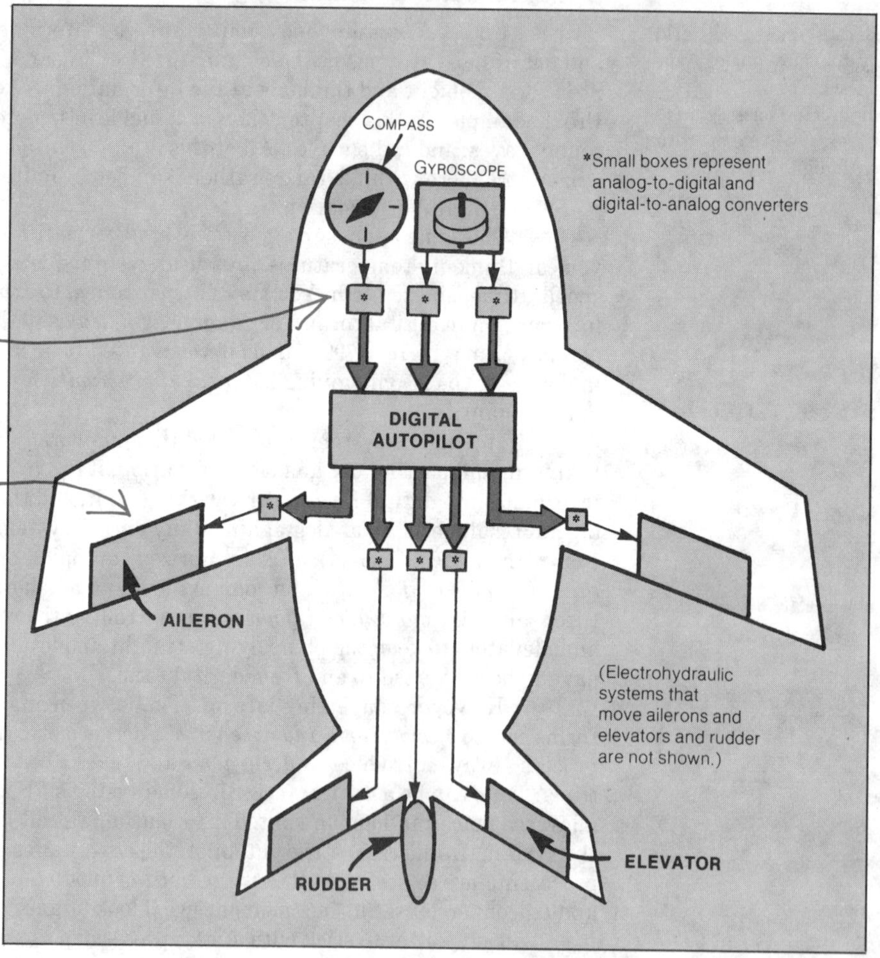

2. Analog processing can be simpler

Okay, suppose we're designing a system that handles analog inputs and outputs as we just discussed. How do we decide whether to process the information by analog or digital methods? In many cases, the answer may be obvious, because we may find that analog processing is much *simpler and more economical*!

Let's look at an example—the phonograph amplifier depicted in *Figure 5-12*. We've got weak *analog* signals carrying sound information from the needle and cartridge (which are *sensing* the information from the surface of the plastic record where it is *stored*). The system's main task is simply to *multiply* the height of these electric waves by a factor depending on the loudness we desire, producing proportionately taller copies of the same waves to drive the loudspeaker.

As we have already discussed, multiplication can be handled with pretty fair accuracy by an analog *amplifier* circuit as shown in *Figure 5-12*. We can make a rather crude but workable amplifier using just *one* transistor, with a few resistors and maybe a capacitor or two. Or as we've mentioned, an IC power amplifier could be used here, very economically.

Trying to accomplish some functions with digital circuits that are normally analog results in a very complicated digital system compared to a simple analog system.

**Figure 5-12.
A Phonograph System is Better as an Analog System**

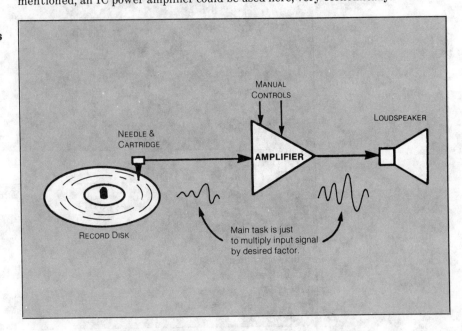

Even a very high-fidelity amplifier would be simpler than a *digital* system to do the same job, as pictured in *Figure 5-13*. This system would check the input voltage regularly every 100 microseconds or so, convert the voltage to a digital number in several wires, multiply the number by a digital volume-control factor (probably from a keyboard as shown, to avoid having to convert an analog signal from a variable resistor), and finally convert the resulting digital product back into an analog output voltage. A new output voltage would appear every 100 microseconds, giving a fair approximation of the taller waves we desire. (Actually, the loudspeaker wouldn't follow the stairstep voltages but would respond to *current* through it, which would get smoothed out somewhat by magnetic effects in the speaker coil.) The "digital-to-analog converter" alone would be considerably more complicated and expensive than a single analog power amplifier.

So *Figure 5-13* shows why you don't see a lot of digital phonograph amplifiers. The same goes for radio receivers and many other fully analog systems. They involve information-processing jobs of a sort that can be handled more economically by analog methods.

**Figure 5-13.
A Digital Phonograph
System**

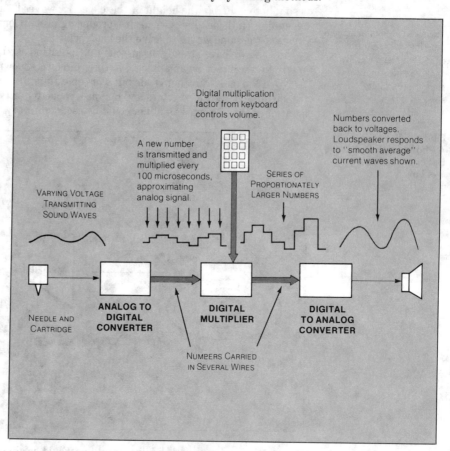

Digital multiplication factor from keyboard controls volume.

Numbers converted back to voltages. Loudspeaker responds to "smooth average" current waves shown.

A new number is transmitted and multiplied every 100 microseconds, approximating analog signal.

SERIES OF PROPORTIONATELY LARGER NUMBERS

VARYING VOLTAGE TRANSMITTING SOUND WAVES

NEEDLE AND CARTRIDGE

ANALOG TO DIGITAL CONVERTER

DIGITAL MULTIPLIER

DIGITAL TO ANALOG CONVERTER

NUMBERS CARRIED IN SEVERAL WIRES

3. Analog systems can transmit information faster

There's one more limitation of digital methods, one that crops up in digital *communications* systems. And that's the fact that when you've got a particular transmission system (counting amplifiers, antennas, wires, or whatever), you can actually transmit information *faster* in the form of analog signals than with digital signals. (You can transmit more information per second.) This limitation only comes into play when you're pushing the capabilities of the transmission system to the utmost, when you're trying to cram as much information as possible through it in the shortest time possible.

Signal transmission systems are limited by the bandwidth of frequencies (the lowest to the highest frequency) that can be handled by a particular band or channel.

To begin seeing why analog transmission is faster, look at the example in *Figure 5-14*. Here, we're transmitting voltage-analog television (video) signals from a remote surveillance camera to a monitor. For the sake of simplicity, we're assuming that the information-handling capabilities of the system are limited only by the wire between the units. Let's say that the wire can't carry variations in voltage occurring any more often than five million analog waves (or digital pulses) per second. That is, the *frequency limitation* of the system is five "megahertz." Furthermore, because the wire is long and not perfectly shielded from outside interference, the voltage signals·may be *inaccurate* by as much as 1/128 of the full range of voltage.

Any transmission system is limited in these same two ways. A communications engineer would say we have a "bandwidth" of five megahertz, and a "signal-to-noise ratio" of about 42 "decibels" (which is an equivalent indication of accuracy). Our wire is just barely capable of carrying a decent television picture in both these respects.

Figure 5-14. Analog Transmission System

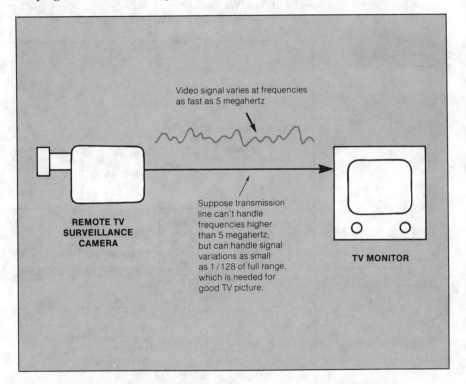

Video signal varies at frequencies as fast as 5 megahertz

REMOTE TV SURVEILLANCE CAMERA

Suppose transmission line can't handle frequencies higher than 5 megahertz, but can handle signal variations as small as 1 / 128 of full range, which is needed for good TV picture.

TV MONITOR

Now since we're talking about speed of information transmission, let's consider just *how much information per second* is being carried by our 5-megahertz, 42-decibel line. The answer will be in terms of *bits per second*, because as we noted back in Chapter 1, the "bit" is the basic unit of information. But how can this be, since bits are *digital* units, and we're talking about an *analog* signal?

Well, we have to figure out how many bits per second *would* be required, *if* we were to transmit the very *same* information in the most efficient *digital code*, which is binary numbers. To do this, we have to imagine a system as shown in *Figure 5-15*.

Here, the analog signal from the TV camera goes to a unit we're calling an "analog-to-digital converter with serial output." Without looking inside this unit, we'll just say that it *measures* the input voltage *ten million times* each second, and converts it into a *seven-bit* binary number. These bits are fed through the transmission line in *series*, as digital pulses with a frequency of *70 megahertz*. At the other end of the line, a "digital-to-analog converter with serial input" puts out a continuous analog voltage signal to the TV monitor. This voltage is proportional to the seven-bit binary number that the converter last received. The broken line below the transmission line in the figure is to remind us that if we actually *built* a system like this, we couldn't depend on the two converters to stay synchronized with each other in handling the serial groups. We would need some sort of *common clock pulses* supplied to both systems, as we learned in studying *Figure 4-4*.

Digital system transmission of analog data, may, in many cases, require much greater bandwidth to accurately transmit the same information.

**Figure 5-15.
Digital Transmission
System**

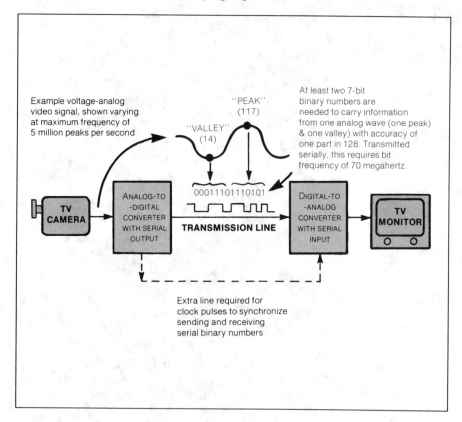

The main point here is that it would require digital signals at *70 million bits per second* to reproduce the same signals with the same accuracy as for the 5 megahertz analog signal. The waveform shown above the transmission line in *Figure 5-15* illustrates *why* this is so—*why we need ten million 7-bit numbers per second* to carry the information from a 5-megahertz analog signal with accuracy of one part out of 128. Ten million numbers per second will give us a good chance of measuring not only the "peak" of each 5-megahertz voltage wave, but also the "valley" next to it. We need to have *both* measurements in order to reconstruct the signal at the receiving end. And seven bits per number gives us a range from zero (0000000) to 127 (1111111). So each measurement is accurate to within 1/128 of the voltage range, which is the accuracy we said was involved in the original analog signal. Therefore, a 7-bit number every 10 millionth of a second results in our 70-megabit-per-second data rate.

The same transmission link that transmitted particular analog information may not be able to be used for digital transmission of the same information because it would not pass the required frequencies.

Now the original transmission line, because of its frequency limitation, could handle only *five million digital pulses per second. It could not handle the equivalent 70-megahertz digital signal.* The signal would lose its amplitude as it passed through the cable and could not be recognized at the other end. Therefore, the information would be lost. With a 5 megahertz bandwidth, the highest frequency information that could be digitized and transmitted is 0.357 megahertz or *14 times slower* than the analog signal capability.

Our explanation of why digital transmission is slower was highly simplified. But the general principles you've seen apply to any *"channel"* that can carry either digital or analog information. This includes telephone wires, radio broadcasts, and microwave radio beams to and from satellites and space vehicles. As we said earlier, this speed or "bandwidth" problem of digital transmission only comes into the picture when we're pushing a transmission system to its limits. But in those cases, it can knock digital methods right out of consideration, as it did in the case of our television system.

However, there is an *advantage* of digital transmission hidden in this situation. Suppose you've got a very "noisy" transmission "medium" (think of a single wire, as we have been doing), but no limit on the bandwidth (frequency). You can send analog signals as *accurate and noise-free as you like* if you put them in *digital* form. To transmit digital pulses accurately, you only need enough accuracy (freedom from noise) to tell a "1" from a "0" at the receiving end. This shows you how the digital advantage of *precision* that we discussed earlier (Number 2 in *Figure 5-9*) applies to the field of *communications*.

Why do some systems use both digital and analog parts?

In this chapter, beside showing you the alternative to digital methods, we have surveyed quite a range of different kinds of electronic systems. So now you can see why we have digital computers and calculators—and other sorts of digital systems—but not digital phonographs and radios and what-not. The reasons are based on a handful of advantages and limitations of digital methods by comparison with analog methods.

But more than that, you have seen that many kinds of systems use *both* digital and analog techniques in various parts of the systems. Indeed, the most important issue that must be settled early in the design of most systems is which parts will use digital methods and which parts will use analog. In some cases the answer may be so obvious as to be a foregone conclusion. But in many cases the answer may depend on careful economic analysis of the trade-offs.

However—as time goes on, the economic benefits of integrated circuits are being applied more and more strongly to *digital* circuitry. So the balance in the choice between digital and analog is shifting further and further toward the digital side. As you proceed through this book, you will come to appreciate more fully the marvelous things that can be done with the simple little switching circuits with which we are becoming familiar.

Quiz for Chapter 5

1. How are analog systems different from digital systems?
   a. They handle information in analog form.
   b. They vary electricity continuously with respect to time and frequency rather than switching it.
   c. They don't use transistors.
   d. A and B above.

2. At any one instant in time, the current through the earpiece of a telephone represents:
   a. Milliamps of current.
   b. A complete sound wave.
   c. Air pressure making up part of a sound wave.
   d. Digital information.

3. What is the proper word for the height or strength of a wave?
   a. Frequency.
   b. Modulation.
   c. Analog.
   d. Amplitude.

4. Which mathematical operation is performed by a "modulator" amplifier?
   a. Addition.
   b. Subtraction.
   c. Multiplication.
   d. Division

5. In an AM or FM system, the waves that are modulated are called the "carrier" waves, as opposed to the "information" waves that modulate the carrier. Would you say the carrier frequency must always be higher than the information frequency?
   a. Yes.
   b. Yes, sometimes.
   c. No, never.
   d. It doesn't matter.

6. Why does frequency modulation represent an analog transmission method rather than digital?
   a. The voltage varies smoothly over a wide range rather than being switched.
   b. The carrier frequency varies smoothly over a wide range rather than jumping suddenly between two definite frequencies.
   c. Both of the above.
   d. None of the above.

7. Why is a capacitor not a perfect way to store an analog voltage level?
   a. It will only store definite levels of voltage.
   b. It's really a digital device, since it's switched on and off.
   c. It only delays information rather than storing it.
   d. Reading out the voltage changes it.

8. Why do we say that a transistor is an *amplifying* device when it is being used as an electrically-controlled variable resistor?
   a. Because it's varying the output rather than switching it.

b. Because it's varying current in response to a voltage control signal.

c. Because the circuit it's used in is called an amplifier.

d. Because a small change in the control signal can produce a larger, "amplified" change in the output current.

9. Why are digital circuits simpler and easier to design with than analog circuits?
   a. They are always made in the form of integrated circuits.
   b. There are more different kinds of circuits to choose from in the digital area.
   c. They are all within the same "family," using the same basic style for the gates.
   d. The circuits don't have to control the electricity precisely over a wide range.

10. Why aren't analog methods used for handling extremely precise information?
    a. Precise information always involves numbers, which are inherently digital.
    b. Analog information never needs to be very precise.
    c. There are limits to how closely an analog signal can reproduce the information.
    d. A and B above.

11. How would an analog system store information for a long time with no change?
    a. By storing voltages in capacitors.
    b. By using delay lines.
    c. By converting the information to digital form and using digital storage techniques.
    d. Any of the above.

12. How fast can a digital gate be made to respond to an input signal?
    a. A few billionths of a second.
    b. A few nanoseconds.
    c. About a second.
    d. A and B above.

13. What results from the fact that inductors, high-capacitance capacitors, and high-precision resistors cannot be economically made in integrated circuits?
    a. Analog circuits are not as extensively integrated as digital circuits.
    b. Digital circuits are not as extensively integrated as analog circuits.
    c. Analog circuits have to force electricity to match outside information.
    d. Digital circuits have to force electricity to match outside information.

(Answers in back of the book)

Digital Integrated Circuits

From the beginning of this book, we've said a number of times that the main reason digital electronics has become so widespread is the fact that digital circuits can be *completely integrated*. We've mentioned a few examples of the fantastic results of using integrated-circuit construction for digital electronics. We've talked about thousands of transistors and other devices mass-produced in a chip a quarter of an inch square, at low cost. We've spoken of gates no bigger than a flyspeck, and so on.

But although you saw a photograph of an IC chip in Chapter 1 and learned how an MOS transistor works in an IC in Chapter 2, we haven't yet looked into how integrated circuits are manufactured. This is a very important subject for our understanding of digital electronics, and not just because it's a very interesting story. The way integrated circuits are fabricated has a lot to do with the differences between the various "families" of digital integrated circuits. In fact, we've reached a point in our learning where we can't go much further without seeing how integrated circuits are made.

So let's find out about what's inside an integrated circuit and how it's put there. This will show us how a calculator held in our hands can contain enough circuitry to fill a closet if each transistor were separate, or an entire room if the circuits used vacuum tubes. Although we can't cover this subject in any great depth, still we can give you a general idea how it's done. Then later on, we'll get around to comparing the various families of digital integrated circuits.

How is an IC chip made as part of a "slice"?

The foundation of integrated circuit manufacturing is a wafer-like, thin, circular slice of crystalline silicon crystal. All circuitry is built upon this silicon slice.

As indicated in *Figure 6-1b*, a hundred or more integrated-circuit chips are made together as sections of a round disk of nearly pure, crystalline silicon called a "slice." Each slice is about three inches in diameter and about 10 mils thick* (10 thousandths of an inch – about the thickness of five to ten sheets of paper). We won't go into how the slices are originally prepared (*Figure 6-1a*) or how the processed slices are cut up into separate IC chips and mounted into packages (*Figure 6-1c* and *6-1d*).

The combination of thousands of P and N regions arranged in a particular pattern on a chip's silicon base produce the circuits which enable the IC to function. The P and N material is created by adding dopants to the silicon.

What's important for us is how thousands of tiny n-type and p-type regions are created on one side of a silicon slice (*Figure 6-1b*) to make semiconductor devices such as the MOS transistor we saw in *Figure 2-11*, and how these devices are connected together to form circuits. (Look again at *Figure 2-11*. Remember, the p-regions and n-regions have different electrical properties, due to small amounts of other substances added to the silicon in these areas. We didn't say so before, but adding these other substances is called "doping" the silicon. And the substances used are called "dopants.")

**Figure 6-1.
Steps in Making an IC**

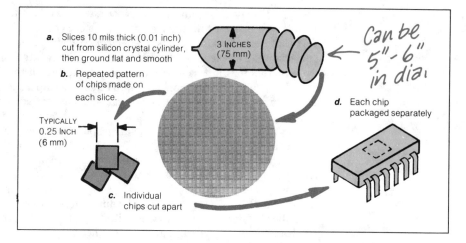

a. Slices 10 mils thick (0.01 inch) cut from silicon crystal cylinder, then ground flat and smooth

3 INCHES (75 mm)

Can be 5"-6" in dia.

b. Repeated pattern of chips made on each slice.

TYPICALLY 0.25 INCH (6 mm)

d. Each chip packaged separately

c. Individual chips cut apart

After the oxide has been deposited, the slice is covered with light sensitive photoresist which is selectively exposed to light through a photomask. The patterns on the photomask form the patterns that are to be the N regions.

Figure 6-2 pretty well summarizes the most important steps in creating these little regions in a slice. (The specific details here are for an n-channel MOS integrated circuit as in *Figure 2-11*.) To begin with, the slices are p-type silicon, because of a particular dopant substance (typically boron) added to the silicon before it crystallized from a hot liquid. The slices are ground flat and polished to a mirror finish. Then, as shown in *Figure 6-2a*, the slices are heated in an oven containing oxygen and steam, causing a very thin layer of silicon oxide to form over the whole surface of the slice where the circuitry is to be created. The oxide film is about 0.04 to 0.08 mil thick. (Remember, a mil is a thousandth of an inch. We are dividing that into 100 smaller parts and speaking of only four to eight of those parts. This is about 1 to 2 micrometres.)

Then, as shown in *Figure 6-2b*, a thin film of light-sensitive liquid plastic called *"photoresist"* is applied to the surface of each slice. The liquid is then dried to form a solid film.

Next (*Figure 6-2c*), the coated slice is exposed to ultraviolet light, except in thousands of tiny spots and strips where n-regions are to be formed. To do this, the light is passed through a glass microfilm plate that covers the slice, called a *"photomask."* The photomask contains a photographically reduced pattern of very accurately-placed opaque (dark) areas to shade the desired parts of the slice from the light. Where the light strikes, it quickly causes the photoresist to become very strong and tough.

Where the N regions are to be located, the photoresist is first washed away. Then a hole is etched by acid through the now exposed oxide to the silicon base.

Then, as shown in *Figure 6-2d*, the soft, unhardened areas of photoresist are washed away with a solvent, like rinsing spots of jam from a breakfast plate. (The hardened areas are left like dried egg yolk on your plate.) Afterwards, the slice is dipped in an acid bath that etches (dissolves) through the oxide film where it's unprotected by the hardened plastic. The acid is a type that doesn't bother the plastic or the silicon, but it eats away the oxide film. A typical hole through the oxide might be about one mil or 25 micrometres across. (The vertical scale in the drawings is stretched about ten times taller than the horizontal scale.)

**Figure 6-2.
Creating an MOS IC**

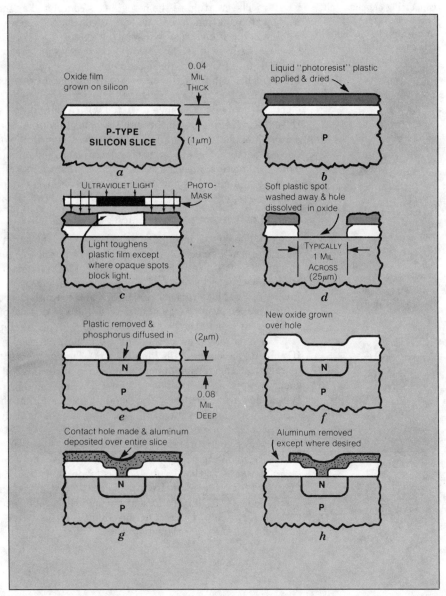

The remaining plastic is removed and the slice is exposed to a high temperature gas. The phosphorus in the gas creates N-type silicon in the areas exposed by the hole pattern. After the N regions are covered with a layer of oxide for insulation, the connecting aluminum electrical conductors are formed by the photomask process.

Next (*Figure 6-2e*), the hardened plastic is removed by a stronger solvent than before. Then the slice is placed in a very hot oven (around 2,000 degrees Fahrenheit, 1,200°C), where it's exposed to a gas containing a substance (typically phosphorus) used as a dopant to make the silicon n-type. The phosphorus diffuses (soaks) into the silicon under each hole through the oxide, changing the silicon from p-type to n-type. The oxide blocks diffusion to all other areas. When the phosphorus has diffused to a depth of about 0.08 mil or 2 micrometres (for n-channel transistors), the slice is removed from the oven and the diffusion stops.

The creation of the n-regions is completed as shown in *Figure 6-2f* by an oxidation process as before, covering the bare silicon spots with a fresh layer of electrically insulating oxide.

As we will see later (*Figure 6-7*), some integrated circuits such as the TTL family require more than one of these cycles of diffusion through different holes in the oxide film, creating *layers* of n and p-regions. Finally, one last set of holes is made through the oxide as before, for electrical contacts. The slice is coated with a thin film of aluminum over the oxide, reaching down through the holes to contact the silicon (*Figure 6-2g*). Then photoresist and acid are used one more time to leave the desired pattern of metal strips for electrical conductors (*Figure 6-2h*). These conductors form the circuit "wires" that we've talked about. There, in a nutshell, is how hundreds of integrated circuit chips are processed together by the same steps as part of the same slice.

How do patterns of p and n-regions make a circuit?

The patterns in which the N and P regions are arranged allow them to fur tion as transistors and other components of an electronic circuit.

The question remains as to how p-regions and n-regions in an IC chip can form an electronic circuit. For an example, *Figure 6-3* answers this question with respect to the n-channel MOS inverter circuit we studied back in *Figure 2-14*. The upper part of *Figure 6-3* shows how the inverter would look (greatly simplified and magnified about 500 times) if it were somehow cut away and lifted out of the surface of the IC chip. This way, we can see all the parts below the surface of the chip.

You should be aware that this drawing represents a region in an IC chip only about six mils long (150µm). That's just a little bit more than the diameter of an average human hair, which is about 4 mils. The tiny specks representing grains in the metal strips would be about the size of a typical bacterium or germ. You could fit six or eight of these inverters into the period at the end of this sentence. So this is truly a microscopically small circuit!

For your reference, below in *Figure 6-3* is a slightly rearranged and more complete version of the schematic diagram from *Figure 2-14*, laid out exactly as the circuit above is made. As is customary in schematic diagrams for integrated circuits, the circles are left off the transistor symbols. The bold lines represent parts of the circuit made of metal on the surface of the chip. The rest of the circuit is either n-type or p-type silicon, as labeled on the schematic diagram.

**Figure 6-3.
An N-channel IC MOS
Inverter**

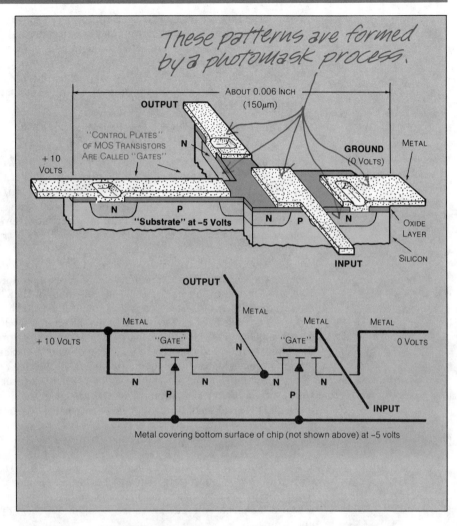

Where are the MOS transistors in this structure?

The load transistor is formed by a long, narrow P region between two N regions. The input transistor's P region is short and broad. The control plate (gate) of the input transistor is connected to an input signal.

If you compare the drawing above in *Figure 6-3* with the MOS transistor we studied back in *Figures 2-11* and *2-12*, you can recognize the load transistor on the left, formed by two n-regions with a long, narrow p-region between them. (It's long from left to right, and narrow in the other direction.) Over the oxide above the transistor is what we called the "control plate" in Chapter 2. The customary name for such a control plate in any MOS transistor is "gate." (We didn't mention this name earlier, to avoid confusion with gate *circuits*.) The gate extends off to the left as a conductor strip for the 10-volt power supply. It's also connected through a hole in the oxide layer to the n-region at the extreme left. (Though the drawing doesn't show it, the oxide over the p regions under the gates is only about a tenth as thick as the oxide elsewhere.)

The input transistor to the right has a channel area between its two n-type terminals that's shorter from left to right and wider in the other direction. Its gate is connected to the input strip, and its right-hand n-type terminal is connected to a "ground" strip at zero volts, through a hole in the oxide. Its left n-type terminal (in the center of the drawing) is the *same n-region* as the right terminal of the load transistor. The output strip is connected to this n-region.

What keeps the electric current in the right places?

Because it is very difficult for current to pass from a N to a P region, and because the voltage on the P region is not strong enough to force current into the N region, current is carried by only the N-regions and metal areas.

As you can see, the entire slice and all the chips cut from it consists of a single large p-region (called the "substrate," meaning "under-layer"), with thousands of little n-regions all over the surface. Electric current is carried only in the n-regions and metal, and very little current ever gets into the p-type substrate. This is because, as we noted in Chapter 2, current (positive charge) cannot easily flow across the junction from an n-region to a p-region, even when the voltage is trying to force it that way.

Current can't pass the *other* way either (*into* the n-regions), because the substrate is not at a high enough voltage to force the current this way. In fact, in this particular kind of n-channel MOS integrated circuit, the substrate is kept at about *negative* five volts, by means of a connection to the back surface of the chip (shown in the schematic diagram below in *Figure 6-3*). This voltage is one way to prevent a certain troublesome interaction between the p-region and the oxide layer that tends to keep n-channel transistors turned on all the time.

So there you have the general idea of how an actual circuit can be formed in an IC chip. The circuit works as though each device (transistors and others we will see later) were in separate packages connected by wires.

What are the advantages and limitations of integration?

Some of the reasons for making a circuit in integrated form (rather than wiring separate devices together) are obvious. You get a lot more circuitry in a much smaller package, and the cost is drastically reduced due to mass-production in large volumes. Other advantages are not so obvious.

For one thing, the miniature circuitry typically requires *less power* to process a given amount of information. Beside saving on power-supply costs, a system doesn't require so much *cooling*. (After all, the power used comes out in the form of *heat*.) Even more important, integrated circuits are much more *reliable*, because the circuitry is contained in a small, tight bundle, so there are no wires to flop around and break loose.

Integrated circuits offer the advantages of small size, low cost, low power consumption, minimal cooling requirements, and high reliability. However, integrated circuits cannot handle large amounts of power.

What an integrated circuit *cannot* do, however, is handle a *great deal of voltage or current*. The p-n junctions can't hold back more than a few dozen volts at most (usually much less). And currents greater than a few dozen milliamps (thousandths of an ampere) generate more *heat* than can be conducted out of the tiny spaces involved, causing temperatures to rise beyond permissible limits. Furthermore, as we learned in Chapter 5, certain electric or electronic devices such as transformers *cannot be made* in an IC chip.

For these reasons, ICs are mainly used for low-power *information-processing*, using circuit designs that don't require many of the "impossible" devices. Where these devices are required, and where high voltage or current are needed (such as for outputs from a system), separate or "discrete" devices are wired into the system, or complete circuits made of such devices.

On what basis are digital IC families compared?

Now that you've got the general picture of how amazingly tiny p-regions and n-regions are created and interconnected to form circuits at the surface of an IC chip, we can proceed to compare the most popular families of digital integrated circuits. These families differ mainly in the types of transistors they use, and consequently in the circuit configuration of the logic gates. These differences give each family certain advantages and limitations in terms of performance and economy. So before we look at some specific families, let's get familiar with the performance characteristics by which a circuit family is evaluated, and what makes one family more economical than another. (See *Figure 6-4* for a summary.)

**Figure 6-4.
Desirable Features of a
Digital IC Family**

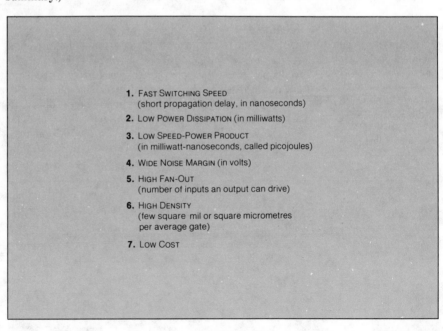

1. FAST SWITCHING SPEED
   (short propagation delay, in nanoseconds)

2. LOW POWER DISSIPATION (in milliwatts)

3. LOW SPEED-POWER PRODUCT
   (in milliwatt-nanoseconds, called picojoules)

4. WIDE NOISE MARGIN (in volts)

5. HIGH FAN-OUT
   (number of inputs an output can drive)

6. HIGH DENSITY
   (few square mil or square micrometres
   per average gate)

7. LOW COST

1. Switching Speed

Propagation delay is a
measure of the speed with
which a gate reacts at its
output to a change at its
input.

The "speed" at which a gate switches (changes its output from
"high" to "low" or vice versa) is measured in terms of "propagation
delay." This is the time the gate takes, after an input is quickly changed
from one state to the other, to make a resulting change at the output.
Propagation delays are usually expressed in nanoseconds (billionths of a
second), or sometimes in microseconds (millionths of a second). In most
applications, we want propagation delays of gates to be as short as
possible, so we can cram digital signals through at a high frequency (bits
per second) and process a great deal of information in a short time.

2. Power Dissipation

All operating electrical circuits generate *heat* in proportion to
both the currents and voltages involved. We refer to this process as the
"dissipation" (meaning waste and scattering) of electrical energy or
power. The speed of heat generation by a gate circuit is usually expressed
in milliwatts (thousandths of a watt), similar to the wattage rating of an
electric heater.

We want the power dissipation of any integrated circuit to be as
small as possible, but not just to conserve electrical power. The heat has to
be *removed,* usually by air flowing around or through the system, or else
the IC chips in their packages would get too hot to operate properly. So
usually the main trouble with power dissipation is that it can *prevent
putting a lot of circuitry in a small space.* This is because the more closely
you crowd the circuits together on a chip, the more difficult it is for the
heat to get out of the silicon to the air, so the circuits may become too hot
to work properly.

3. Speed-Power Product

The ability of a circuit to
remove or dissipate its
heat influences how closely
gates may be arranged.
The speed-power product
is a measure of a circuit's
power dissipation and
propagation delay.

Due to the nature of transistor switching circuits, usually *the
shorter a gate's propagation delay, the more power it dissipates.* To see how
well a particular gate circuit achieves *both* short propagation delay and
low power dissipation, we look at its "speed-power product." This is just
the propagation delay multiplied by the power dissipation. It is usually
expressed in units of nanoseconds times milliwatts, a combination called
"picojoules." ("Pico" is an abbreviation for 10^{-12}, meaning 0.000000000001,
and "joule" is a certain amount of energy.)

4. Noise Margin

The noise margin of a gate
should be as wide as possible to prevent electrical
noise voltages from changing its logic state.

The noise margin of a gate or any other digital building-block tells how securely it transmits and receives information without errors, in spite of "noise." Noise in this case means unwanted, "stray" voltage signals picked up by the various wires connected to the circuit. The main design feature that's important here is the difference between the voltage *transmitted* for each "logic state" and the voltage *required at an input* to be accepted correctly as one of the logic states. If the transmitted high and low voltages are 1 volt higher and lower than would be barely necessary to be correctly recognized, the noise margin is 1 volt. This is because it would take a noise signal greater than 1 volt to cause an error in such a system. Obviously, we want a system to have as *wide* a noise margin as possible.

5. Fan-out

The greater a gate's ability
to supply or sink current
to or from other gates on
its output, while at the
same time requiring mini-
mum current on its input,
the better.

The fan-out of a certain gate in a network is the number of *inputs to other gates* that its output is connected to. You want gates with a *high fan-out capability* – gates that can drive a lot of other gates. This is another way of saying that you want gates that can *supply (or accept) lots of current* at their outputs but *don't require* much current at their inputs (either momentary current to change states or continuous current to maintain a state). Loading an output with too many inputs reduces the noise margin and speed of the gates involved.

6. Density of Circuitry

You want gates that occupy a *small area on the IC chip*, so you can pack lots of gates on each chip and thus make the circuitry very dense (or "complex"). The area a gate occupies on the average is measured in square mils (thousandths of an inch) or in square micrometres. The density of an IC depends not only on the factors we've mentioned before (the size of the transistors and other devices, how many are required per gate, and the number of inputs to each gate), but also on the *power dissipation*. This is because devices that dissipate a lot of heat can't be too small or too close together or they would get too hot.

7. Manufacturing Cost

The larger the number of
manufacturing steps re-
quired to produce an IC,
the more expensive it is.
Larger chips do not cost
less to make, and can actually be more expensive because a larger area of a
slice is scrapped by a defect which ruins the
circuit.

The cost to make a standard, mass-produced IC chip depends mainly on its *size* and the *number of processing steps* required to produce the circuitry on the chip. Each processing step (especially each photomasking cycle) costs a certain amount, of course. But more than that, each step tends to introduce a certain number of *defects* on each slice, due mostly to foreign particles and scratches on the photomask. A single speck of dust on a photomask casts a shadow on the photoresist perhaps as big as a transistor, possibly making an entire chip worthless.

Obviously, the bigger a chip is, the fewer you can make in one slice for the same cost. But worse than that, the greater is the chance of at least one defect occurring to spoil any given chip. So *the bigger the chips are, the more chips you have to scrap,* and the more the remaining good ones cost. This *cost factor* sets an upper limit to how big you can make a chip in the effort to get more gates in the same package.

What are the main families of digital integrated circuits?

Finally now, we've learned enough about digital integrated circuits to understand the differences between the most important of the many different families of digital ICs. Keeping in mind the fabrication methods and desirable characteristics we've studied, let's take a brief look at the main "logic" families. ("Logic" is a shorter name than "digital integrated circuit." You'll remember this use of the word "logic" from Chapter 4).

1. P-Channel MOS Logic

P-channel MOS logic integrated circuits have a N-type substrate with P-type diffusions, which is the reverse of the N-channel circuit discussed earlier. Most P-channel devices require a negative voltage supply and are designed to operate on the principles of negative logic.

There are about a dozen different varieties of integrated circuits that use MOS transistors of various sorts, including more than one way to make n-channel circuits such as those we've studied since Chapter 2. Some n-channel circuits provide the best propagation delay and density of all MOS ICs (about 20 nanoseconds and 9 square mils or 5,800 square micrometres per gate).

To give you an idea of the variety among MOS ICs, let's look at a very similar (and somewhat more economical) type of MOS circuit called "p-channel." *Figure 6-5* shows a p-channel MOS 4-input negative NOR gate. An entire p-channel MOS integrated circuit consists of circuits much like this.

As you can see on the left in *Figure 6-5*, the positions of n-type and p-type silicon are simply the *opposite* from an n-channel circuit such as in *Figure 6-3*. (Being a 4-input gate, this circuit has *four* input transistors in parallel as we saw in *Figure 2-17* for a 2-input gate.) We have an *n-type substrate* with *p-type diffusions* into it. P-channel enhancement-mode MOS transistors work just like the n-channel type, except that they are turned on by a *negative* ("minus") voltage signal to the gate metal.

Furthermore, the load transistor must be connected to a more negative voltage supply than the input transistors (minus 5 volts in this case). And finally, we need a separate voltage supply for the load-transistor gate (minus 17 volts). This is because the minus five volts that we're using for the main power supply isn't enough to keep this transistor turned on properly. (Plus or minus five volts is a standard voltage, so that two or more circuit families can be used together easily.)

As you can see from the schematic diagram and partial function table on the right and below in *Figure 6-5*, all inputs must be "high" (about 0 volts) to produce a "low" output (about minus five volts), by turning off all four input transistors. Thus, this circuit is a positive NAND gate. Most p-channel logic designs are prepared on the basis of *negative* logic, so this circuit is usually regarded as a *negative NOR gate*. (Incidentally, note that the little arrowhead in the p-channel transistor symbol points the *opposite direction* from the one in the n-channel symbol. This is how you tell the difference in a circuit diagram.)

**Figure 6-5.
IC Layout of a P-channel
4-input NAND Gate**

All inputs must be high for the output to be low.

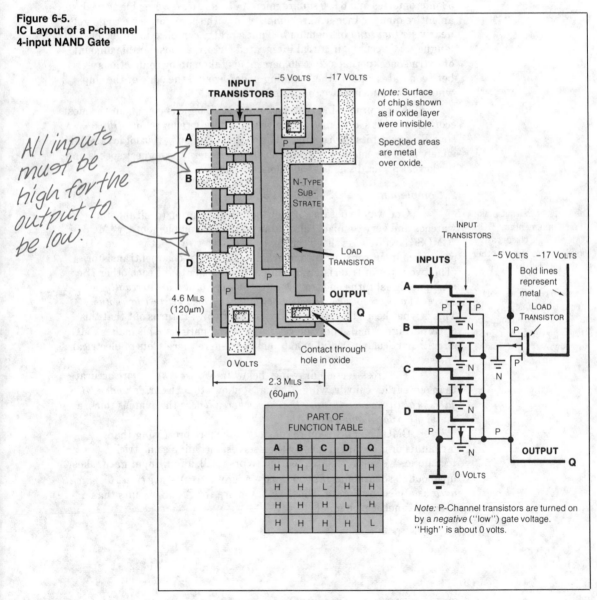

PART OF FUNCTION TABLE				
A	B	C	D	Q
H	H	L	L	H
H	H	L	H	H
H	H	H	L	H
H	H	H	H	L

The performance of any IC depends on factors such as design, size and shape of transistors, voltage requirements, and amount of doping. The P-channel MOS logic IC is very economical.

The performance qualities of a p-channel circuit, like those of any integrated circuit, depend on many factors apart from the general design principles. Such factors include the size and shape of the transistors, the voltages used, and how heavily the silicon is doped. Typically, however, the gate of *Figure 6-5* would have a propagation delay of 40 nanoseconds, a power dissipation of 0.8 milliwatt, a speed-power product of 32 picojoules, and a noise margin of 0.5 volt when used with a fan-out of ten.

As far as circuit density is concerned, this gate (being about 4.6 mils by 2.3 mils) occupies about 10.6 square mils. (That's about 116 by 58 micrometres and 6,800 square micrometres.) If we were able to pack an entire quarter-inch-square (6mm) chip with these gates, we could get nearly *six thousand* of them on it. That's pretty complex, in anybody's language! Of course, in a real integrated circuit, a considerable amount of extra space (perhaps 30 to 40 percent) is taken up by connections between gates, and by the "bonding pads" around the edge of the chip where external wires are connected.

As for processing cost, p-channel MOS logic is just about the most economical that's available. As we've seen, it only requires *one diffusion step*. Because of its high complexity and good economy, this logic family is widely used for complex chips in calculators, where the highest speeds and the lowest power dissipation are not required.

2. Complementary MOS (CMOS) Logic

The transistors that make up a complementary MOS (CMOS) logic circuit are complements of each other (when one is turned on the other is turned off). Because current flows only during the transition when the transistors change states, power dissipation is very low.

One way to obtain very low power dissipation (desirable for watches and very compact calculators) is to use "complementary" MOS or "CMOS" circuitry. As shown in *Figure 6-6* for a two-input positive NOR gate, each input controls both a p-channel and an n-channel transistor. (The two types of transistor are called "complements" of each other.) So an input signal turns one transistor on and the other off. *Virtually no current flows except when an output switches from one state to another.* This is what keeps the power dissipation very low. It turns out that this circuit design can also provide very high noise margins. This makes CMOS circuits particularly useful in automobiles, where there's lots of electrical "noise."

The cross-section drawing (below in *Figure 6-6*) is more accurate than our earlier chip drawings, in that it shows that the *oxide under MOS gates is always much thinner than elsewhere.* Where the oxide is thick, a metal strip *does not act as a gate.*

CMOS circuits require considerably more processing than p-channel or n-channel types by themselves, including some tricky techniques we don't need to mention. Worse still, a four-input gate takes up about 50 square mils (32,000 square micrometres) — *five times as much area* as a p-channel gate. Thus, we pay more for CMOS circuits than for n-channel or p-channel.

Figure 6-6.
CMOS 2-input NOR Gate

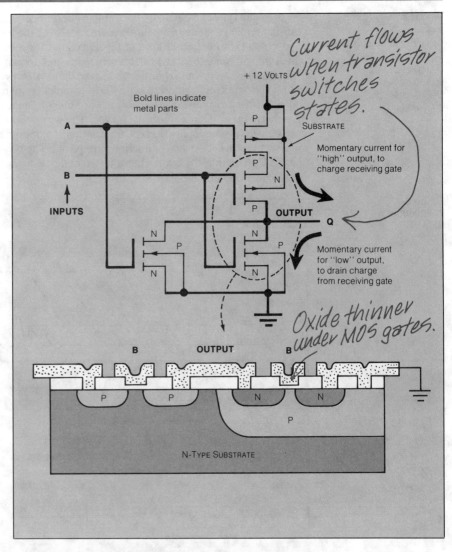

3. Transistor-Transistor Logic (TTL)

Transistor - transistor logic (TTL) IC's are commonly used where a short propagation delay is required. The bipolar transistors used in TTL logic do not require as high a supply voltage to operate as MOS transistors.

The propagation delay of MOS gates is limited by the fact that turning MOS transistors on and off requires *moving a considerable amount of charge* in and out of the metal gates, through some fairly high resistances. This is a relatively slow process. To attain shorter propagation delays, we have to use logic circuits made from "bipolar" transistors rather than MOS transistors. By far the most commonly used, best all-around logic family for a wide variety of general applications is a large family of *bipolar* integrated circuits called "transistor-transistor logic," or TTL. *Figure 6-7* summarizes the idea of a TTL gate and shows what a bipolar transistor is.

Bipolar transistors are made in a N-P-N structure. The emitter N region is doped much heavier than the collector N region. The base P region is the input for the controlling current.

　　　　The type of bipolar transistor used here is called an "n-p-n" transistor, shown structurally and schematically at the lower right in *Figure 6-7*. (The only other kind of bipolar transistor is called "p-n-p," which we will come to later.) In this structural picture, an n-p-n transistor *looks* somewhat like an n-channel MOS transistor. However, it's best for the n-type "emitter" region to be an "n + (n-plus)" region, meaning it's "doped" with an *extra-large amount* of the "dopant" material (typically phosphorus) that makes silicon be n-type. The n-type "collector" region needs much *lighter* doping, except for another n + region next to the metal contact where good conduction is required. The p-region is called the "base." Unlike an MOS transistor, a metal and oxide sandwich plays no part in the operation of a bipolar transistor.

**Figure 6-7.
IC Layout of a TTL NAND
Gate**

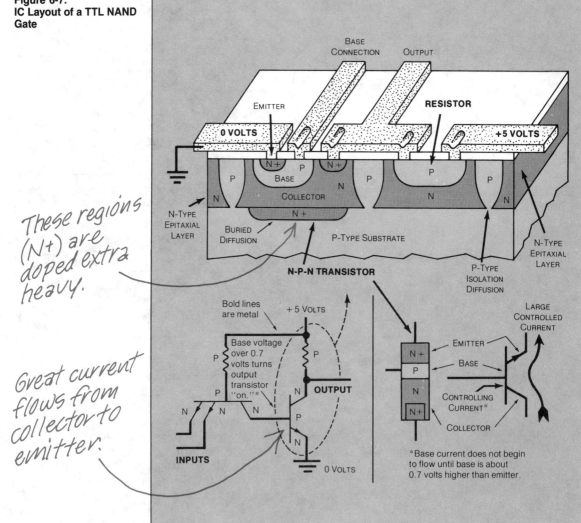

These regions (N+) are doped extra heavy.

Great current flows from collector to emitter.

A small current into the base controls a large emitter-collector current, thus the transistor acts like a current controlled switch. A bipolar transistor can turn on and off more quickly and can carry a larger current than a comparable MOS transistor.

As shown at the transistor symbol at the lower right in *Figure 6-7*, an n-p-n transistor is turned on by making the base voltage at least 0.7 volt higher than the emitter voltage, and then supplying a small current to the base region. This allows a much greater current to flow through from collector to emitter, in proportion to the base current. Thus, the transistor acts as a *current-controlled switch* (or variable resistor, if it's not turned all the way on or off.) The main advantage of bipolar over MOS transistors, as far as we're concerned, is that they can be made to switch on and off much *more rapidly*. They can also carry much more current than an MOS transistor of the same size, and they have much lower resistance when switched "on."

There are several varieties of TTL circuits, but the one basic feature they all have in common is the two-transistor arrangement shown at the lower left of *Figure 6-7*, which is a positive NAND gate. These two n-p-n transistors give the family its name: transistor-transistor logic. As you can see, one transistor has a *separate emitter region* for each input. This means two separate n-regions in contact with the *same p-type base region*. We'll see how this circuit works in a moment.

First, however, notice the physical structure of a typical transistor and resistor in a TTL chip, shown above in *Figure 6-7*. To make a TTL slice, we start with a p-type substrate, and use photomasking and diffusion to create some very heavily-doped n-plus regions. These n + regions (only one is shown) will wind up being *buried* beneath the collector region of each n-p-n transistor, providing low-resistance paths to help these regions conduct electricity better. Then the oxide is removed and a layer of lightly-doped n-type silicon crystal is grown over the surface of the entire slice, from a hot gas containing silicon and phosphorus. This process, and the layer, are called "epitaxial."

The photomask patterns and dopants used, and the order and number of manufacturing steps are significantly different for TTL circuits as compared to MOS circuits.

Then three more diffusions are made. First, a very deep p-diffusion is made all the way down to the p-type substrate, at all places where there will *not* be any devices for the circuit. This is to *isolate* (insulate) a *separate n-region for each device*. It provides a sort of "pocket," using the fact that current can't pass from n to p. Next, a shallower p-diffusion is made into spots in each n-region, for transistor bases, resistors, and the "p" parts of "p-n diodes" (devices not shown here).

Finally, another heavy n + diffusion is made into spots in the base regions to provide emitters, and in the collector regions to provide good electrical contact points. Afterwards, the desired pattern of metal interconnections is produced.

Obviously, this TTL process is a complicated and expensive one, with many opportunities for at least one fatal defect to occur in any chip on the slice. Even so, TTL is remarkably economical, with very good performance. This is what makes it the most popular single logic family.

Figure 6-8 summarizes the way a basic TTL two-input positive NAND gate works, showing (by means of arrows) where the current flows in two different typical situations. In essence, the multiemitter transistor performs the positive-AND function, and the "output" transistor inverts the result and adds power.

**Figure 6-8.
Current Paths in a NAND
TTL Gate**

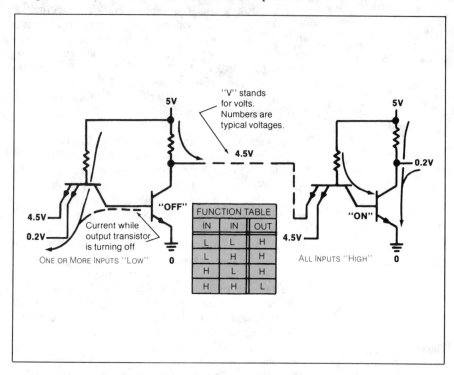

FUNCTION TABLE		
IN	IN	OUT
L	L	H
L	H	H
H	L	H
H	H	L

All inputs to the TTL positive NAND gate must be high for the output transistor to turn on. On the other hand, a low input signal on any input will result in the output transistor being turned off.

As indicated for the gate on the left, *any one input signal* lower than about 0.6 volts (here, 0.2 volts) withdraws current and thus prevents the output transistor from turning on, so that output current at about 4.5 volts is supplied through the resistor. When the voltage at *all inputs* is higher than about 0.8 volts as shown for the gate on the right, current supplied to the base of the multiemitter transistor turns the output transistor on, so that the output draws in current at about 0.2 volts.

This particular style of TTL gate in *Figure 6-8* is intended only for use *inside* an integrated circuit. *Figure 6-9* shows one way in which several such gates can be used inside an IC for maximum complexity and density. Several positive NAND gates *share the same resistor* connected to their outputs, giving us the effect of several AND gates feeding a NOR gate. Each "partial" gate is called an "expander gate."

Below in *Figure 6-9* is shown a typical layout for a four-input NAND expander gate. (The expanders in the schematic drawing above have only two inputs, to keep the drawing simple.) Note that the long, narrow resistor takes up nearly *half the area* of the entire expander gate, which is about 20 square mils in all. Obviously, *resistors are a problem* in efforts to design densely-packed, complex circuitry.

**Figure 6-9.
TTL NAND "Expander"
Gates**

Several partial gates, called expander gates, share a common output resistor. A problem with expander gates is that this common resistor covers almost fifty percent of the total circuit area.

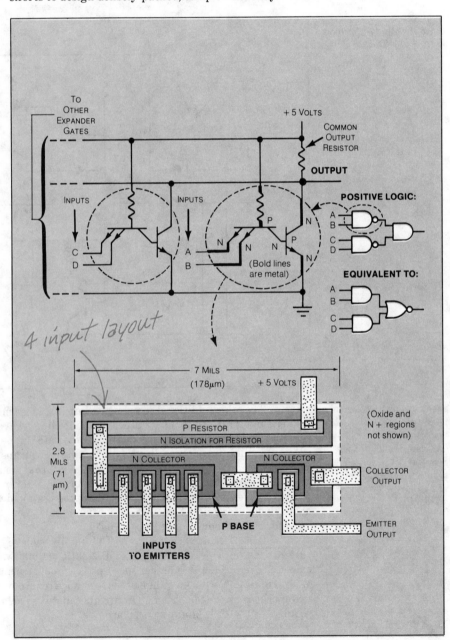

Figure 6-10 shows a more advanced variety of TTL-type NAND gate, in a sub-family called "54/74 low-power Schottky TTL," This is a very popular form of TTL originated by Texas Instruments. The circuitry in the outlined area is the "TTL" part, although the multiemitter transistor is replaced by special "Schottky diodes" to reduce input current for the "high" level.

**Figure 6-10.
Schottky-Diode-Clamped
2-input NAND Gate**

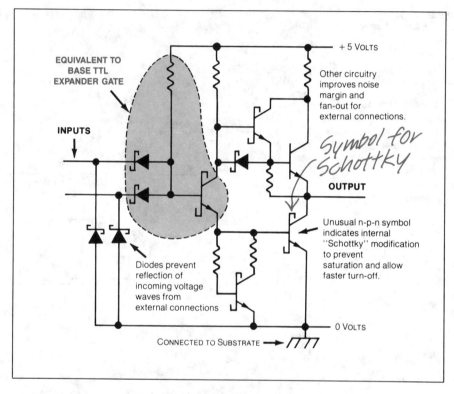

The other additional components improve noise margin and fan-out, and those at the inputs prevent steep voltage waves from being reflected back from the inputs. As a result, this gate can be connected to external inputs and outputs. Best of all, however, the transistors whose symbols include square "hooks" have a special internal modification called "Schottky diode clamping" that makes them much *quicker to turn off*. This provision gives low-power Schottky gates a propagation delay of only about 10 nanoseconds. The power dissipation is only about 2 milliwatts per gate, for a speed-power product of 20 picojoules.

Other circuit varieties in the TTL family achieve either faster speed (3 nanoseconds) or lower power dissipation (1 milliwatt), at the expense of a considerably higher speed-power product. There are varieties of TTL to suit a very wide range of performance requirements where faster speeds are needed than MOS circuits can give. This is another fact that makes the TTL family so popular.

Schottky, diodes which replace the multiemitter transistor and Schottky diode clamping allow the Schottky TTL transistors to be turned off quickly and to have a low power dissipation.

4. Emitter-Coupled Logic

Emitter-coupled logic (ECL) transistors have an extremely short propagation delay because they never completely turn on. The disadvantage is that they have very high power dissipation.

The only logic family that can provide shorter propagation delays than the fastest Schottky TTL is "emitter-coupled logic," or ECL. *Figure 6-11* shows a typical ECL 2-input positive NOR gate. The transistors and resistors are constructed much like those we saw for TTL in *Figure 6-7*. This type of circuit makes the transistors switch very fast by *never letting them turn "all the way on"* (a condition called "saturation"). The transistors just switch current from one path to another, giving us a propagation delay (for some ECL circuits) of *less than one nanosecond*.

If you like, you can read further explanations in *Figure 6-11*. The important fact to note is that ECL circuits draw *lots of current*, producing a power dissipation well over 40 milliwatts per gate. This heat prevents ECL from being used in large, complex chips, but the high speed makes it useful in many large computers.

**Figure 6-11.
ECL 2-input NOR Gate**

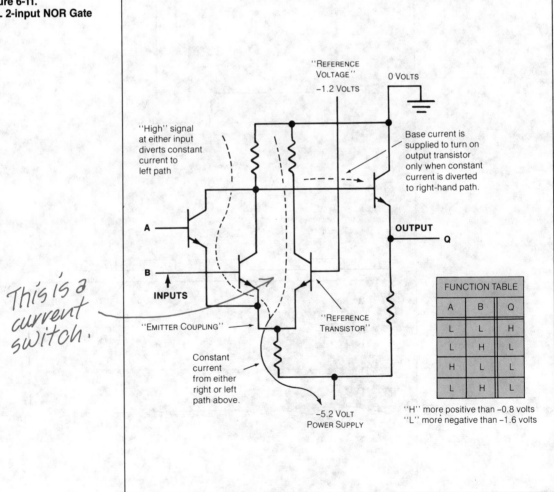

"REFERENCE VOLTAGE"
–1.2 VOLTS

0 VOLTS

"High" signal at either input diverts constant current to left path

Base current is supplied to turn on output transistor only when constant current is diverted to right-hand path.

A

B

INPUTS

"EMITTER COUPLING"

"REFERENCE TRANSISTOR"

Constant current from either right or left path above.

–5.2 VOLT POWER SUPPLY

OUTPUT

Q

This is a current switch.

FUNCTION TABLE		
A	B	Q
L	L	H
L	H	L
H	L	L
L	H	L

"H" more positive than –0.8 volts
"L" more negative than –1.6 volts

5. Integrated Injection Logic (I²L)

The search for a gate circuit requiring the least possible area (and the necessary low power dissipation to go along with it) has led IC designers to a gate with essentially *only one transistor per gate*. This is "integrated injection logic." It's called I²L (I-squared L), as if the initials IIL were an algebra formula. *Figure 6-12* shows a simplified cross-section of an I²L positive NAND gate, with a schematic interpretation.

**Figure 6-12.
An I²L NAND Gate**

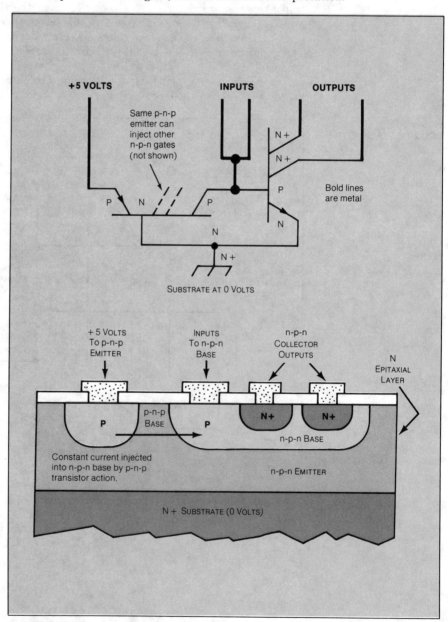

The integrated injection logic (I²L) gets its name from the injection of a constant current from the p-n-p transistor into the base of each n-p-n transistor. An I²L gate requires the least possible area and has the lowest power dissipation of the logic types discussed.

On the right in both drawings in *Figure 6-12* is an n-p-n transistor that is the gate. It's similar to the n-p-n transistor in *Figure 6-7*, but simpler, and *upside-down*. The epitaxial n-layer over the entire n + substrate acts as the *emitter*. (It's also the emitter for all other n-p-n transistors on the chip, so *no isolation diffusion* like that in *Figure 6-7* is needed.) One n + diffusion for *multiple collector regions* is made into desired spots in the p-type diffused base area. Although this structure looks like a multiple-*emitter* transistor, it's *used* as a multiple-*collector* device, and it works in such fashion well enough for this application.

Over to the left is another diffused p-region that runs a long way through the chip (perpendicular to the paper), past several different n-p-n transistors. This p-region, together with the base region of each n-p-n transistor and the n-region in between, act as a multiple-collector *p-n-p transistor*. A p-n-p transistor works just like an n-p-n transistor, with everything in reverse. It's turned on by current being *withdrawn* from the base, and the larger controlled current flows from emitter to collector.

Because of the way its base is always at the zero volts of the substrate, this p-n-p transistor supplies a *constant current* into the base of each n-p-n transistor. This "injection" of current into the n-p-n bases is where we get the name, "integrated injection logic." But how in the world can a single n-p-n transistor with constant base current and multiple collectors act as a positive NAND gate?

Figure 6-13 answers this question. In the schematic diagram here, we see two of these special n-p-n transistors, each one being supplied with base current by a collector of the p-n-p transistor above. Two inputs are connected to the base of each n-p-n device. (We could have more inputs if we wanted.) As indicated by the peculiar two-output NAND symbols (shaded areas), each n-p-n collector acts as a separate output. One limitation of this family is that *each output can be connected to only one input* of a receiving gate. Notice that an output *accepts* current to transmit a "low" signal, but *turns off* to transmit a "high" signal.

The circuit in *Figure 6-13*, which is called an "$\overline{S}$-$\overline{R}$ latch" (S-bar, R-bar), provides an example of how I²L gates transmit to one another. The gate on the left shows what happens when all inputs are "high" (not withdrawing current). All the continuous current injected by the p-n-p transistor above flows into the n-p-n base, turning the transistor on. Thus, both outputs are connected to ground (0 volts) for a "low" signal.

The gate on the right shows what happens when at least one input is "low" (connected to ground). All the injected base current is *diverted out the "low" input* (or inputs, if more than one is "low"). So the transistor is *not* turned on, and the outputs are effectively "high."

Figure 6-13 also shows how this S-R latch might be laid out in an IC chip. Only about 15 square mils (9,700 square micrometres) are required for these two gates and their interconnections. Obviously, I²L gives us very high packing density.

**Figure 6-13.
An I²L Ŝ-Ř Latch**

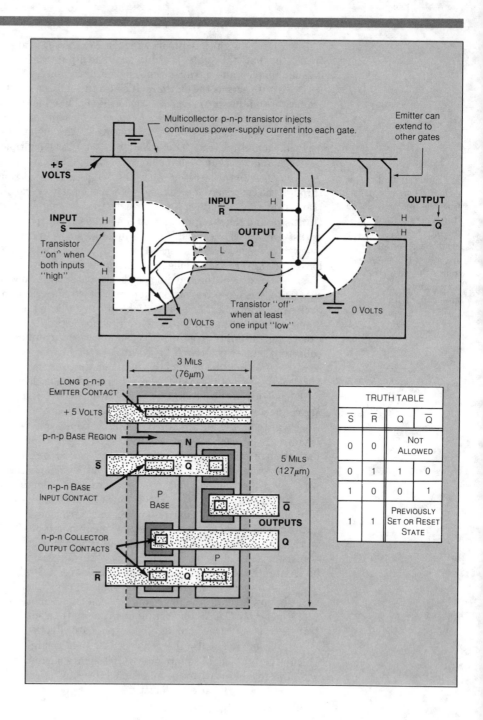

Multicollector p-n-p transistor injects continuous power-supply current into each gate.

Emitter can extend to other gates

+5 VOLTS

INPUT R H

OUTPUT Q̄ H

INPUT S H

OUTPUT Q L   H

Transistor "on" when both inputs "high" H

L

Transistor "off" when at least one input "low"

0 VOLTS

0 VOLTS

3 MILS (76μm)

LONG p-n-p EMITTER CONTACT

+ 5 VOLTS

p-n-p BASE REGION

N

Ŝ   **Q̄**

n-p-n BASE INPUT CONTACT

P BASE

5 MILS (127μm)

Q̄

OUTPUTS

n-p-n COLLECTOR OUTPUT CONTACTS

Q

P

Ř   **Q**

TRUTH TABLE			
Š	**Ř**	Q	Q̄
0	0	Not Allowed	
0	1	1	0
1	0	0	1
1	1	Previously Set or Reset State	

An I²L gate designed for a 20-nanosecond propagation delay will *dissipate only about 0.05 milliwatt,* giving us a speed-power product of only *one picojoule.* So these tiny gates can be packed as closely as we like, without worrying about overheating. They are ideal for watches, which must operate for months on a single small battery. Furthermore, I²L techniques are being rapidly improved, so we are likely to see this family come on very strong in the next few years.

What have we learned about digital integrated circuits?

Once again, we've come a long way in one chapter. We've seen the types of processing steps that can mass-produce circuits too small to be seen without a microscope. We've looked under the surface of IC chips and seen how the components work together to function as logic gates — the building-blocks of which all information-processing circuitry is composed.

This understanding enabled us to survey two MOS logic families and three bipolar families that represent most of the digital integrated circuits manufactured. We've seen that MOS circuits, due to their compact gates and simple processing, provide large, complex, and inexpensive chips. Complementary MOS gives very low power dissipation, at the expense of reduced density and economy. TTL is the most popular all-around type of circuit, with a good balance of all the desirable characteristics. ECL provides the fastest possible gates, at the expense of high power dissipation, which also prevents making large, complex chips. Finally, I²L is an up-and-coming type of circuit that manages by ingenious design to give us a gate consisting essentially of only one transistor, with unbelievably low power dissipation.

Armed with this understanding, we can proceed now to study methods by which digital systems store large quantities of information. Then we can move on to how the parts of entire systems work together.

Quiz for Chapter 6

1. The dense, accurate pattern of tiny dark spots on a transparent photomask is prepared by what method?
   a. Photographic reduction from a larger version, as with microfilm.
   b. Drawing actual size by hand, with tiny ink-pens.
   c. Drawing a bigger version on a plastic sheet and shrinking it in an oven.
   d. None of the above.

2. A speck of dust on a photomask for an n-diffusion can spoil an entire:
   a. Slice, by causing one defect.
   b. Chip, by causing an n-diffused spot.
   c. Chip, by preventing a spot from being diffused.
   d. Photomask.

3. Current is kept flowing in the right places in the silicon by:
   a. Silicon oxide insulation between transistors.
   b. The photoresist layer.
   c. The fact that voltage cannot easily force positive charge from an n-region to a p-region.
   d. All of the above.

4. Generally speaking, the size of a chip and the number of processing steps required determine:
   a. Noise margin.
   b. Fan-out.
   c. Cost.
   d. All of the above.

5. P-channel MOS circuitry provides:
   a. Very short propagation delays.
   b. Large, complex chips.
   c. Low-cost processing.
   d. B and C above.

6. Complementary MOS circuitry is especially strong in the area of:
   a. Dense, complex circuitry.
   b. Low-cost processing.
   c. Low power dissipation.
   d. Very short propagation delays.

7. Which family is the most commonly used, best all-around choice for a wide variety of general applications?
   a. P-channel MOS.
   b. Complementary MOS.
   c. Transistor-transistor logic.
   d. Emitter-coupled logic.

8. Which family provides the shortest propagation delays?
   a. P-channel MOS.
   b. Complementary MOS.
   c. Transistor-transistor logic.
   d. Emitter-coupled logic.

9. What advantages does integrated injection logic offer?
   a. Short propagation delays.
   b. Density and complexity.
   c. Very low power dissipation.
   d. B and C above.

(Answers in back of the book)

Mass Storage in Digital Systems

We began our study of digital electronics with a complete system—the calculator in Chapter 1. Since then, we've narrowed our scope down to individual *parts* found in most digital systems—gates, flip-flops, and building-blocks of various kinds. In due time, we're going to put all our accumulated understanding of the parts together and consider entire *systems* again. But before we do so, there's another kind of *part* that we need to take a look at, which is "mass storage units" in digital systems. So that will be our subject for this chapter.

What is a "mass storage unit"?

The mass storage unit (or memory) of a digital system is able to store a great number of instructions and information for long periods of time.

Another name sometimes used for "mass storage unit" is "mass memory," or sometimes just "memory." We're talking about a sort of "reference file" where relatively *large quantities* of information are stored, perhaps for a relatively *long time*. This general definition sets mass storage units apart from *building-blocks* that store information, such as those we studied in Chapter 4. You wouldn't call a single flip-flop or a counter a "mass storage unit." And a single register would not be considered a mass storage unit unless it could hold a large number of bits, as we will see later.

The difference between *mass* storage unit and *other* units that store information will become more clear to you as we go along. However, it may be helpful at this point to look at some examples in our calculator system, shown in *Figure 7-1*.

Obviously, the microprogram memory subsystem is a mass storage unit. As you will remember from Chapter 1, this unit contains many instructions, stored there when the IC chip was made, that the controller refers to at each step in the calculator's operation.

Any one of the six *registers* shown in *Figure 7-1* would *not* be a mass storage unit *by itself*. However, in a typical calculator, the three "number" registers (and perhaps even the flag register) would be made as *parts of a single storage unit* that *would* qualify as a mass storage unit. It would be a fairly large reference file for information to be put into and taken out of as necessary, and where information could be left for as long a period as desired. These examples from the calculator should give you a general idea of what we're talking about when we say "mass" storage unit or memory.

What's important about mass storage?

The reason mass storage is important enough for us to devote a whole chapter to it is that it involves two subjects that we will refer to again and again in connection with digital systems: *cost* and *speed*.

First of all, mass storage units contribute a *large part of the cost* of most digital systems (as well as a large part of the physical size). To appreciate this fact, take another look at the photograph of the calculator chip in *Figure 1-3*. Notice that the microprogram memory and the number and flag registers take up nearly *half the area of the entire chip*. We know that means half the cost, as well.

Furthermore, the speed with which a system works (how much data it can process in a given time) may be determined by *how fast* information can be moved in and out of mass memory units. So the mass memory units may to a great extent determine not only the *cost* of a digital system but also its *usefulness*.

For these reasons, a great deal of the *progress* in squeezing more and more of the parts of a system into one (or just a few) integrated circuits has resulted from advances in the design of integrated-circuit mass-storage units (memories). Being able to store more bits in smaller areas in an IC chip has made possible one-chip calculators, smaller computers, and many other systems at a much reduced cost. So this is obviously an aspect of digital electronics that we need to look into further.

Mass storage of a digital system usually takes up a large portion of the physical space and a large part of the cost required for the system. The speed of the system also depends greatly on the efficiency of the memory operations.

**Figure 7-1.
Mass Storage Units in the Calculator System (asterisked units)**

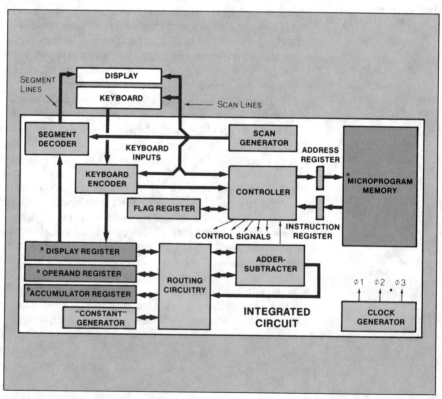

How are mass storage units classified?

Through the years, many different techniques have been tried for storing bits of information in digital systems. Out of all these methods, the ones presently used for mass memories can be classified in several categories, as summarized across the top of *Figure 7-2*. So far, we've only studied the first method listed —flip-flops. We'll concentrate mostly on the methods that show the most promise for the future, which are those using *integrated semiconductor storage elements*.

Notice in *Figure 7-2* that storage-unit types are also classified according to "access method." For now, we'll say that access method means the *order or sequence* in which stored information is retrieved (located and read out) from a memory unit. In a *serial-access* memory, the stored information is available for reading only in a certain order, usually the *same order in which the information was put in*. On the other hand, information can be taken out of storage locations in a *random-access* memory in *any* order "at random."

The difference between these two access methods will become more clear to you as we look at some examples. *The importance* of the difference is due to the fact that random-access-type memories are generally much *faster* at getting information in and out, as we will see.

Mass storage uses semiconductor devices, magnetic devices, or punched hole devices for the storage method, and serial or random access for the access method.

**Figure 7-2.
Classification of Mass
Memory Types**

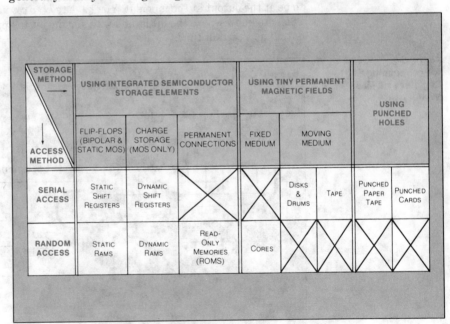

STORAGE METHOD → / ↓ ACCESS METHOD	USING INTEGRATED SEMICONDUCTOR STORAGE ELEMENTS			USING TINY PERMANENT MAGNETIC FIELDS		USING PUNCHED HOLES	
	FLIP-FLOPS (BIPOLAR & STATIC MOS)	CHARGE STORAGE (MOS ONLY)	PERMANENT CONNECTIONS	FIXED MEDIUM	MOVING MEDIUM		
SERIAL ACCESS	STATIC SHIFT REGISTERS	DYNAMIC SHIFT REGISTERS			DISKS & DRUMS / TAPE	PUNCHED PAPER TAPE	PUNCHED CARDS
RANDOM ACCESS	STATIC RAMS	DYNAMIC RAMS	READ-ONLY MEMORIES (ROMS)	CORES			

How can shift registers provide "serial-access" mass memory?

Since shift registers are the only type of mass memory unit in *Figure 7-2* that we're already familiar with, let's begin our study with them. First, let's see how shift registers are typically *made into* mass memory units, and then look at an application in a system.

You may want to refer back to Chapter 4 *(Figure 4-3b)* to remind
yourself of what a shift register is. It's a string of one-bit storage units
with a common clock. (In Chapter 4, the one-bit storage units were
flip-flops. We'll see another kind a little later.) At each clock pulse, the
stored bits *shift* one step along the string.

Now when a shift register is used for mass storage, it is usually
provided with a data selector and a "recirculation path," as shown in
Figure 7-3. The bits in the register are *continually* shifted to the right by
regular clock pulses, without ever pausing. When the "loading control"
input is at "0," each bit that's shifted out the right end of the register is
shifted *right back into the left end.* Thus, the output shows us every bit in
the register, one at a time, over and over again, like a Ferris wheel going
around. Recirculating shift registers can be made in any length. Some
(like the *AMD TMS 3114)* are over a thousand bits long.

Data is loaded into this recirculating memory through the "serial
data input," by switching the loading-control input to "1." A new bit is
stored every time the clock is pulsed.

It's obvious why this is called a "serial-access" memory. It's
because the stored bits appear at the output *in series.* You can't access
(get to) a stored bit immediately to read it or change it, unless it happens
to be at the output at the moment. You have to *keep count of the clock
pulses* to know where the bit is, and wait for the bit to appear at the
output at a certain time.

One type of serial access
memory is provided by a
recirculating shift register.
Access to a specific bit is
gained only when the de-
sired bit shifts in sequence
to the output. Shift registers are very useful for
mass storage in applica-
tions where the informa-
tion naturally flows in
serial form.

**Figure 7-3.
A Recirculating Shift
Register as a Mass
Memory Unit**

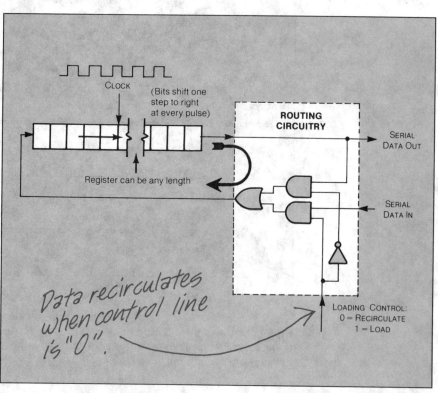

What's a typical use for a serial-access memory?

Serial-access memory units such as recirculating shift registers can be very efficient as mass memories in applications where the data *naturally* needs to go in or out (or both) in *serial form*. A good example is the three number registers in the calculator system of *Figure 7-1*. (Remember, each register by itself would hardly be a mass memory. But all three together serve as the calculator's mass data memory.)

Figure 7-4 shows *why* serial access can be useful in these registers. It comes in handy when the two numbers in the display register and the operand register are being added or subtracted. The binary-coded decimal digits of each number go to the adder-subtracter *one at a time in the order in which they're stored*. First, the two "least-significant digits" (those on the far right) are added or subtracted, then the next digits to the left, and so on.

In the next chapter, we'll see just *how* the four bits of each BCD digit are added or subtracted. (It's somewhat more complicated than adding pure binary numbers, which we learned about in Chapter 3.) For now, we'll just point out that as soon as two digits are sent to the adder-subtracter, the adder-subtracter produces the appropriate "sum" or "difference" digit. These resulting digits must be loaded into the accumulator register in the *same sequential order* (least-significant digit first).

The combined operation of the three number registers of a calculator is an example of a mass memory system that is efficiently served by serial access. This is because the digits are always processed one at a time (sequentially) in the order they are entered.

**Figure 7-4.
Serial Digit Movement in
Number Registers**

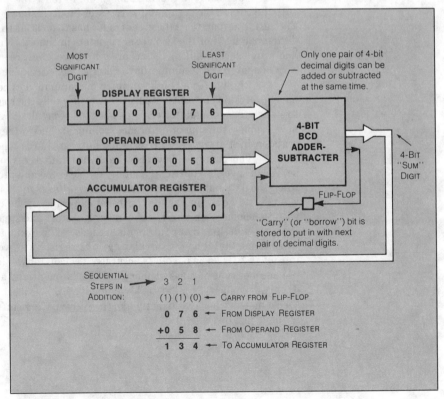

As each pair of digits is added—say, the 6 and 8 shown in *Figure 7-4*—the adder-subtracter also produces a "carry" bit (which for 6 and 8 would be "1," along with a sum bit of 4). The carry bit is stored in a flip-flop and added in with the next most significant pair of digits (in this case, the 7 and 5). Similarly, a "borrow" bit is stored in the flip-flop during subtraction.

Now our so-called mass memory in *Figure 7-4 doesn't have to* be designed with serial access. In fact, there are many calculators that do *not* have a serial-access mass data memory (because they're a type of "microcomputer" *programmed* as calculators). But the point of this example is that it would be very *convenient* for the registers in *Figure 7-4* to have serial access. So let's see how recirculating shift registers could be used in this "naturally serial" application

How can a calculator use recirculating shift registers?

Serial-access memory units must have the mass memory recirculating shift registers synchronized with the controller so that all operations occur in the correct sequence.

As indicated in *Figure 7-5*, we would use *four* recirculating shift registers for each number register—one for each bit (8, 4, 2, 1) in the binary-coded decimal digits. All twelve shift registers are *clocked in step together*, so that the bits for all three least-significant digits show up at the outputs (and the inputs too, of course) at the same time.

When the least-significant digits appear at the outputs, the controller tells the adder-subtracter to begin adding (or subtracting). And the controller tells the routing subsystem to route the adder-subtracter's output digits to the accumulator-register input as shown in *Figure 7-5*. The rest is automatic. All the controller has to do is sit back and watch the digits march out of the two upper registers in time with the clock pulses, and watch each resulting "sum" or "difference" digit roll right into the correct spot in the accumulator register at each clock pulse.

Right after all eight digits in our simplified calculator's numbers have been added or subtracted, the controller tells the routing subsystem to switch the accumulator register back to "recirculate." And we've got the new number (the sum or difference) rolling around in the accumulator register, in step with the two original numbers in the other registers.

This example shows how handy a serial-access memory unit can be. It can save the controller a lot of trouble (time and program steps) when the data naturally needs to move *serially*, as in this example. It works just fine so long as the controller can somehow be *synchronized* with the memory unit, so the controller can start and stop processing with the correct digits. In our future discussions of the example calculator, we will assume that it uses recirculating shift registers as we have just seen in *Figure 7-5*. (To be perfectly accurate, a calculator would more likely use another type of mass memory called a "sequentially accessed dynamic memory," similar to "dynamic RAMS" we'll see later. However, such a memory behaves exactly like the recirculating shift registers that we have just discussed.)

**Figure 7-5.
Shift Registers for Serial
Mass Memory**

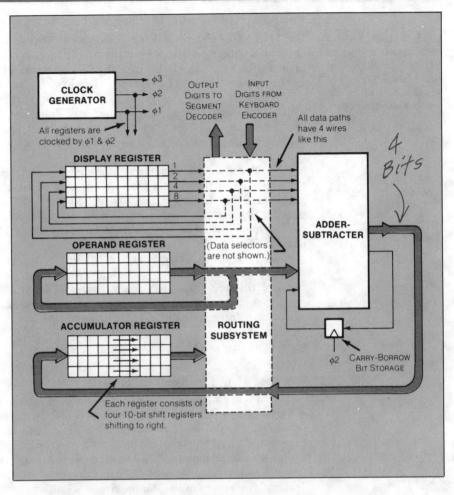

What's another "naturally serial" storage application?

A computer's CRT display also can use a recirculating shift register to store the characters that are displayed on the screen. Each shift register for a CRT line is recirculated in synchronization with a sweep of the light spot.

Before we leave the subject of how recirculating shift registers are used, we should take note of a particularly common application. This is in television-type display units used with computers—units called cathode-ray-tube (CRT) terminals. The rows of characters (letters and numbers) displayed on the screen are typically stored in recirculating shift registers in the terminals, using a seven-bit code for each character. The registers recirculate in step with the "flying spot" of light that paints the TV picture, one horizontal line at a time.

The details are more complicated than we would care to go into. We should merely note that while the picture is being held constant for the viewer to look at, the flying spot repeats the *same sequential pattern* of brightness and darkness as it sweeps across the screen, line by line. So this is another "naturally serial" storage application that recirculating shift registers fit right into.

How are mass-memory shift registers different from others?

Using the storage of electric charge to represent a bit in a dynamic storage circuit is an efficient way to use integrated circuit area to accomplish this.

Shift registers used as mass memories are just like those used as sequential building-blocks, except that they generally store a much *greater number of bits.* (In the case of a CRT terminal, 7,000 bits would be needed to store 25 rows of 40 characters each, with seven bits to identify each character.) Consequently, it's more important that the storage element for each bit occupy the *smallest possible area* on the IC chip. We can't afford any space for extra frills such as parallel inputs or outputs, or shifting backward as well as forward. (This is the reason we have to use *recirculation* for access.)

To reduce the area in which each bit is stored, many mass-memory shift registers use a storage method we haven't talked about yet. So far, the only way we've seen to store a bit in an integrated circuit is in a *flip-flop.* As we learned in *Figure 4-1,* this involves two gates cross-coupled to keep each other in their present states. The *other* way is listed in *Figure 7-2* as "charge storage." It's used not only in mass memories but also in many sequential building-blocks, in the place of flip-flops. This method involves storing each bit as an *electric charge,* in a type of MOS circuit called a *"dynamic storage"* circuit.

As there are many types of flip-flops, there are also many types of dynamic storage circuits—all of them occupying *less chip area* than flip-flops. As an example of a dynamic storage circuit, let's see how a typical *dynamic shift register* works.

How does a typical dynamic shift register work?

Figure 7-6 shows the idea of one of the most common types of dynamic shift register. It consists of a chain of dynamic storage circuits for one bit, like the one in the rectangle. Such a circuit serves the same purpose as a master-slave flip-flop *(Figure 4-5).* Thus, it could also be used in sequential building-blocks such as registers and counters.

As indicated in the clock timing diagram below, the two MOS "gating" transistors in the storage circuit are alternately turned on and then off by the clock signals $\phi 1$ and $\phi 2$ ("ϕ" is the Greek letter "phi," standing for "phase.") When each gating transistor is turned on and off, the voltage signal coming from the inverter to the left charges a *capacitor* on the input lead to the inverter on the right. In this respect, the circuit works like the *analog* storage circuit back in *Figure 5-8.* The capacitor *holds* the voltage level after the gating transistor is turned off. If this voltage is at the "high" level, the inverter to the right transmits a "low" signal, and vice versa.

The MOS dynamic shift register has capacitor-equivalent circuitry on the inputs of the storage transistors to store the charge representing a bit. Since the charge leaks off the "capacitor", the charge must be replenished at a minimum rate to prevent loss of information.

There's no way to prevent a stored charge from leaking. But so long as the voltage is still recognizable as definitely a "high" or "low" signal, the inverter transmits a *strong* copy of it (in inverted form) to the next charge-storage unit. There, the inverted signal is locked in by the *next* clock pulse and strongly inverted back to its original form. Thus, each bit is stored as a regularly renewed charge being passed from circuit to circuit in the shift register.

The main *disadvantage* of dynamic shift registers is that they *must not be left standing still*. They must typically be clocked faster than about 100 hertz (100 full clock cycles per second—a $\phi 1$ pulse and a $\phi 2$ pulse.) Otherwise, the stored voltages will decay too far to be recognized correctly by the inverters. But of course, that's no problem in a recirculating memory, which is kept moving all the time anyway—usually much faster than that.

The need for frequent renewal of charge levels is what gives us the name "dynamic" for this type of storage circuit. "Dynamic" implies energy *in motion*. On the other hand, "static" means "stationary." A *flip-flop* is called a *static* storage unit because it can hold a bit in one place for any length of time.

**Figure 7-6.
A Typical MOS Dynamic
Shift Register**

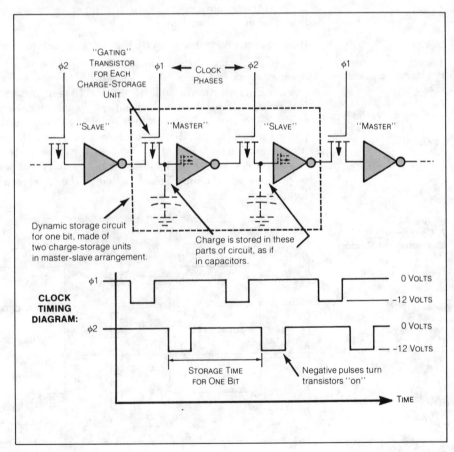

How do dynamic shift registers compare with static shift registers?

Not only do MOS dynamic shift registers take up much less space and cost less than MOS static shift registers, but they also can be clocked much faster.

If you'll compare the master-slave dynamic storage circuit in *Figure 7-6* with the more complicated master-slave flip-flop in *Figure 4-5*, you'll see why dynamic storage units occupy *less area on a chip*. As we noted earlier, this is why dynamic shift registers cost less per bit of storage capacity than static shift registers.

Dynamic shift registers have another advantage, in that they can be clocked *faster than MOS static shift registers* (perhaps 5 megahertz compared to 2.5 megahertz. Remember, a megahertz is a million times per second). When even faster shifting is required, we've got to use *bipolar* circuitry (flip-flops), at a much greater cost per bit.

Later in the chapter, we'll see another kind of dynamic storage unit. But for now, let's look back at *Figure 7-2* for a moment. We've covered the two types of serial-access memories that use integrated semiconductor storage elements (static and dynamic shift registers). So let's move on to learn about mass memories with *random access*. We'll begin with the category of "read-only memories" or ROMs. (ROM rhymes with "Tom.") Here again, we're on fairly familiar territory, since the *microprogram memory* in our calculator *(Figure 7-1)* would be a ROM. So let's use the example of the microprogram memory as a jumping-off point into random-access memories.

How does the microprogram ROM illustrate random access?

The information in a ROM is stored in the circuit during its manufacture and cannot be changed. It can be read, but new information cannot be stored (written).

First, let's recall from Chapter 1 what the microprogram memory does, as shown in *Figure 7-7*. To begin with, the controller puts an "address" number in the address register. In response, the memory unit locates an instruction (stored when the chip was made) in a place identified by that address within the memory. A copy of the instruction is put in the instruction register for the controller to use. Note that *any* instruction can be "read" at any time, in any order. So this is a *random-access* rather than a serial-access memory, but one that can *only be read*, not written into. You can't change the data stored at any memory location.

**Figure 7-7.
A Calculator
Microprogram Memory**

Random Access Memory that can only be read not written into.

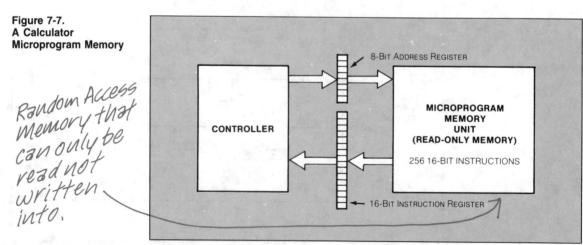

Both the addresses and the instructions are in *binary* form, consisting of ones and zeros. A typical calculator might have 256 instructions, each consisting of 16 bits, for a total of 4,096 stored bits. To count addresses from zero (00000000) to 255 (11111111), each address must be eight bits long.

To begin understanding how such a memory works, let's first see the general way in which the storage units are arranged in *all* random-access memories—both the "read-only" types and others we'll come to later.

How are storage units arranged for random access?

A word is a group of related bits (such as the bits that represent a particular number) that are stored together in and read together from a random-access memory (RAM).

Figure 7-8 shows the general idea of how one-bit storage units (or "cells") of *any sort* would typically be arranged so that stored information can be read out at random. (The same arrangement works for *writing* too, as we'll see later.) For the sake of simple explanation, this memory stores only 16 bits, as eight "words" of two bits each. A "word" is a group of bits that are *stored* together in a random-access memory, and also *processed* together when possible. (For example, each 16-bit instruction stored by the microprogram memory in *Figure 7-7* is a word.)

**Figure 7-8.
A Simple Random-
Access Memory**

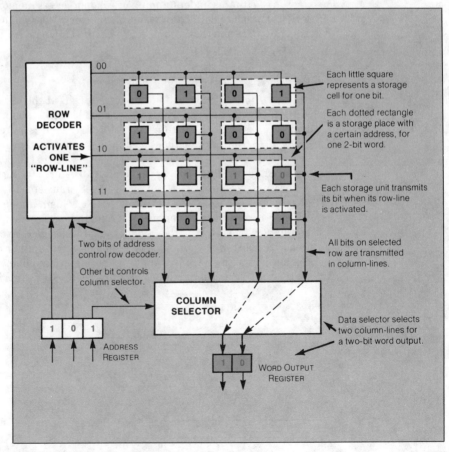

ROW
DECODER

ACTIVATES
ONE →
"ROW-LINE"

00
01
10
11

Each little square
represents a storage
cell for one bit.

Each dotted rectangle
is a storage place with
a certain address, for
one 2-bit word.

Each storage unit transmits
its bit when its row-line
is activated.

All bits on selected
row are transmitted
in column-lines.

Two bits of address
control row decoder.

Other bit controls
column selector.

COLUMN
SELECTOR

Data selector selects
two column-lines for
a two-bit word output.

1 0 1

ADDRESS
REGISTER

1 0
WORD OUTPUT
REGISTER

The one-bit storage cells in *Figure 7-8* are arranged on the chip in a square pattern of four horizontal "rows" and four vertical "columns." Each cell is connected to one of four horizontal "row-lines" and one of four vertical "column-lines." The two cells for a given word are on the *same row line*. (By comparison, the calculator ROM would have 64 rows, each storing four 16-bit words.)

The RAM usually uses a rectangular array or matrix formed by a pattern of horizontal rows and vertical columns. A word's address selects a particular row and a particular column to locate the word.

Addressing eight words requires *three-bit* address numbers, running from zero to seven. (In *Figure 7-8*, we're addressing word number five—101). Two of the address bits go to a "row decoder," causing it to activate one of the four row-lines (in the figure, it's row number two, or binary 10). This makes all four cells in that row transmit their bits in their column lines. Of these four bits, two (one word) are selected by the "column selector." This is a data selector controlled by the remaining bit of the address (1, in the example). Thus, the addressed word (10) is placed in the "word output register."

Nearly all random-access-type memories use one variation or another of this basic idea of a rectangular "array" (arrangement) of one-bit storage units. It allows the storage units to be packed closely together, without an excessive amount of interconnecting wires.

With this understanding of random-access memories in general, let's look in more detail at a read-only memory. This will be a particular kind of ROM—one that's made in a p-channel MOS integrated circuit, such as a calculator chip.

How does an MOS read-only memory work?

Figure 7-9 shows four bit-storage cells at the lower left corner of a ROM array of any size. Each cell (shaded blocks) is simply one MOS transistor—or rather, it is if a "1" is stored. If a "0" is stored, the cell is an *incomplete* transistor, without a gate, so it can *never be turned on*.

The row decoder transmits a "low" voltage (negative 5 volts) in the selected row line. This *turns on* all transistors storing "1" on that row, connecting a column line through each MOS transistor to a "ground line" at zero volts. Thus, a "1" is transmitted in a column line to the column selector as zero volts. Otherwise, if a column has an incomplete transistor on this row (meaning 0), the column is kept at *negative five volts* by a "load" transistor down below. (This should sound familiar to you. Each column-line with its transistors, neighboring ground line, and load transistor acts as a many-input *negative NOR gate* as in *Figure 6-5*.)

**Figure 7-9.
P-channel MOS Read-
Only Memory Cells**

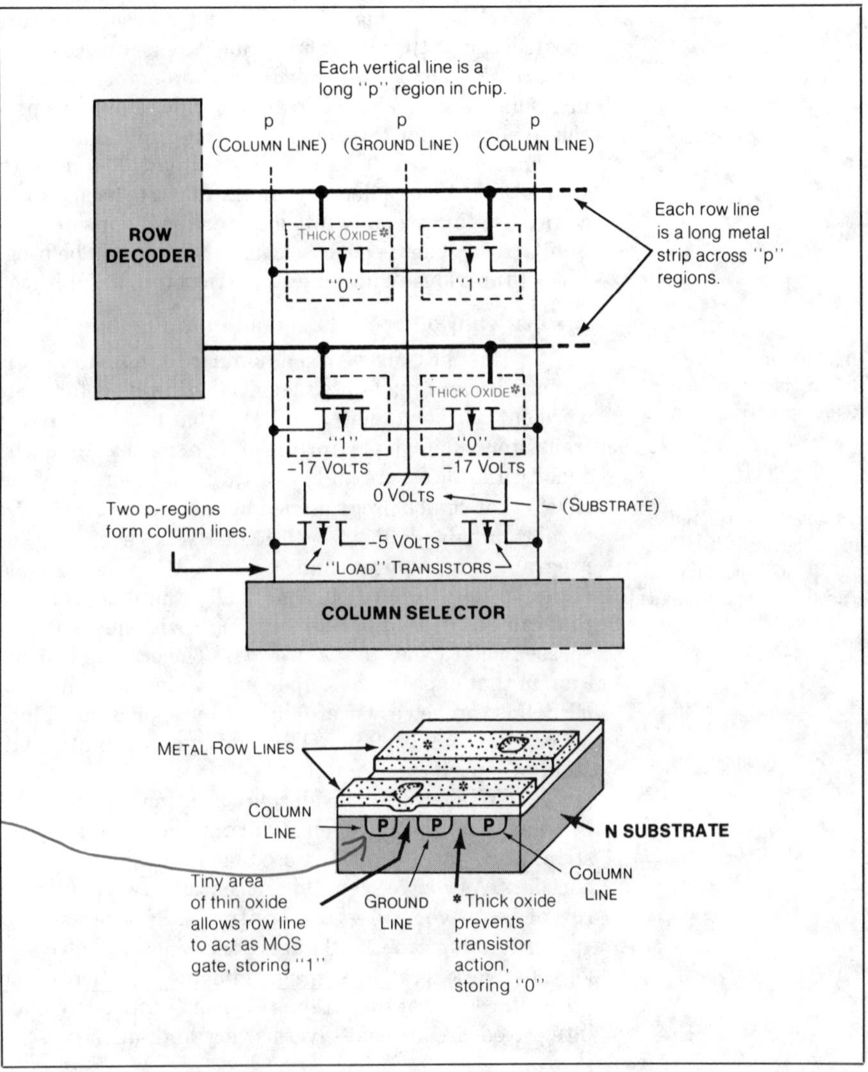

The P type material is laid out into long columns like the metal row line.

In a MOS ROM, the row lines are long strips of metal laid over the oxide; the column lines and ground lines are long strips of P-type material running crossways under the row lines. Whether a 1 or a 0 is stored at a particular location is determined by the thickness of the oxide under the metal.

Further below in *Figure 7-9*, you'll see how this array of cells is built with very high density (low cost per bit!) in an IC chip. Each row line is simply a long metal strip over the oxide layer. Each column line and ground line is a long p-region running crossways under all the row lines. Where a transistor is desired (for permanently storing a "1"), the oxide is made *very thin* under the metal where it crosses over the silicon between a column line and a ground line. Where the oxide is *thicker*, the electric field from the metal is too far from the silicon to "turn on" a channel as we discussed with regard to *Figure 2-12* for n-channel transistors. (Remember, we mentioned thin oxide for the first time in discussing *Figure 6-6*.)

Beside its high density, the most useful feature of this ROM construction is that the "thin-oxide" areas are created in only *one step of photomasking and etching of oxide*. (Remember, we discussed photomasking back at *Figure 6-2*, although not this particular step.) So it's relatively easy for the factory to "program" a customer's specified information into a ROM integrated circuit. The ROM then becomes a special product supplied to that customer, typically in rather large quantities for assembly into a system built for sale by the customer. Similarly, a general-purpose calculator chip can be programmed for special functions required by a particular calculator model.

Are there any other kinds of read-only memories?

In addition to MOS ROMS, bipolar-based ROMs are available with increased access speed, but at a higher cost. ROMs that can be programmed after manufacture are called PROMS; and ROMs that can be erased and reprogrammed are called EPROMs.

MOS ROMs with cells similar to those in *Figure 7-9* are the most commonly-used read-only memories. However, for faster access (getting a word out quicker), there are ROMs that use *bipolar* rather than MOS transistors—but we pay for the higher speed with much lower packing density and therefore higher cost. Furthermore, there are certain MOS ROMs that can be programmed by the user after they're made, called "programmable ROMs," or "PROMs." However, this "do-it-yourself" programming requires connecting the IC to a special electronic system designed for this purpose. And finally, there are some kinds of PROMs that can be *erased and reprogrammed* with new data. These are called "eraseable" PROMs, or "EPROMs." The erasing is done in a special system that exposes the chip to ultraviolet light. Although PROMs and EPROMs can be very useful in certain applications, their bit density is lower than that of plain ROMs, and they cost quite a bit more per bit of storage capacity.

While we're considering variations on the ROM idea, you should be aware that the high-density arrangement of gate strips and diffused strips in *Figure 7-9* is often used in *building-blocks* such as decoders and encoders. As you've seen, this is basically a way to make a close-packed row of negative NOR gates (positive NAND gates). Because of the ease of "programming" the MOS gate connections as we've seen, a ROM used as a logic network is called a "programmable logic array," or "PLA" (pronounced letter by letter as "P-L-A"). PLAs are one feature of MOS integrated circuits that gives us very high-density, economical IC chips.

How are random-access memories different from ROMs?

Although ROMs are accessed in a random fashion as we've seen, they're not *called* "random-access memories." That name is reserved for memories that can be *written into* as well as read from. The name is abbreviated to "RAM," pronounced like the word for a male sheep. Later on, we'll consider how RAMs are used in electronic systems. But for now, let's find out how RAMs are made and how they work.

As we learned earlier in the case of shift registers, there are also two general types of RAM: dynamic RAMS that store bits in the form of electric charges, and static RAMs that store bits in flip-flops. Let's look first at *dynamic* RAMs.

How do dynamic RAMs work?

Dynamic RAMs fit the general pattern for all random-access-type memories that we discussed in *Figure 7-8.* (Remember, we said you'd see those general features in *all* memories with random access.) Electric charges are put into the cells through the column lines and *read out through the same lines,* using appropriate switching circuitry in the "column selector" section. As in all dynamic storage units, the stored charges decay in a fraction of a second, so they have to be "refreshed" often by one method or another.

Figure 7-10 shows the general idea of one of many types of dynamic RAM. Due to the exceedingly tiny size of each storage cell, this one integrated circuit (perhaps 0.16 inch or four millimetres square) can store 16,384 bits in 128 rows and 128 columns.

In a read-write RAM, information can be written into the circuit as well as read out. The addressing of a read-write RAM works the same as a ROM.

Figure 7-10.
A Dynamic RAM One-Transistor N-channel Cell

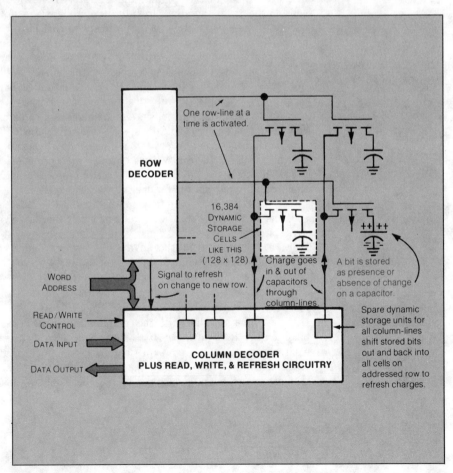

Each cell consists of one n-channel MOS transistor and a tiny capacitor—which as we noted earlier is a device that stores electric charge at a certain voltage. Each capacitor is simply a small area of metal over the oxide. When a row-line is activated (with a higher voltage), all the n-channel transistors on that row are turned on, connecting their capacitors to their column lines. By way of the column lines, the capacitors are charged when "writing," and the charges are detected when "reading."

The capacitors (each made from a metal plate over the oxide) that are contained in the RAM cell hold the charges that store the data in RAM. Refresh circuitry and refresh cycles are required to keep the capacitors charged. Thus, the name dynamic RAM.

We won't go further into electrical details. Suffice it to say that words from the "data input" in *Figure 7-10* are written into the capacitors as electric charges by voltage signals through the column lines, and read out the "data output" through the same column lines. The "read-write control" input tells the subsystem whether to read or write. Such a memory can be designed to handle words of any length (from one to 128 bits in this example), so we're not specifying word length in the figure.

Every time a row-line is activated, the charges in all cells on that row are *refreshed* automatically by a "spare" dynamic storage unit shown at the lower end of each column line. The stored bits are *shifted* from the cells on the row being accessed, into the spare storage units, and right back into the cells again, with renewed strength. This process resembles the shifting of bits in the dynamic shift register in *Figure 7-6*. Any reading or writing that needs to be done takes place at the same time, for selected columns. (As we've noted before, the reason the stored charges need to be refreshed is to prevent the stored data from being lost due to leakage that changes the voltages.)

To make sure all cells are refreshed often enough, the system controller (not shown) typically stops its work for about 50 microseconds (millionths of a second) every two milliseconds (thousandths of a second). During this time, the controller addresses one word on each of the 128 rows. This triggers the automatic refreshing process that we spoke of in the preceding paragraph, for each row addressed.

To summarize the features of dynamic RAMs, they use very small cells—as small as one transistor and one capacitor, as we've seen. This gives us very high packing density, and therefore low cost per bit of storage. However, to get this advantage, we have had to add the additional refresh circuits and program the system's controller to do a "refresh" cycle several hundred times a second.

How do static random-access memories work?

If we can't live with the complication of the refresh cycles, we have to use a RAM of the *static* variety. Static RAMs use a *flip-flop* for each storage cell, in the row-and-column arrangement that we've become familiar with. The flip-flops can be made either of bipolar or MOS transistors. As you might expect, MOS gives us higher density (lower cost per bit) but lower access speed.

As with any kind of RAM, the challenge in designing a static RAM is to make a storage cell that occupies a *small area* and requires very *few electrical connections*. Let's look at just one of many different kinds of static RAM cells, called a "diode-coupled" bipolar cell. This will give us an idea of just *how simple a flip-flop can be.*

The general idea of a bipolar RAM with diode-coupled cells is shown in *Figure 7-11*. Like certain other RAMs, this one has *two column-connections to each cell,* called "bit-lines." As indicated in the logic diagram of one cell *(Figure 7-11a),* each bit-line is connected to the *output* of one inverter and across to the *input* of another inverter. (The two "diodes" act as one-way valves, as explained further in *Figure 7-11c.* Thus, they prevent all the cells in one column from interacting with each other.)

The row-line connection to each cell provides a *ground connection* for both inverters. (A positive power-supply voltage is also provided to each inverter, as shown.) As you can see in Figure 7-11b, this is a pretty simple, compact circuit. It contains just two transistors, two resistors, and the two diodes.

Flip-flops arranged in a row-column matrix provide the storage mechanism for static RAM cells. These do not require charge refreshing. The transistors may be bipolar or MOS.

**Figure 7-11.
A Bipolar Static RAM Cell**

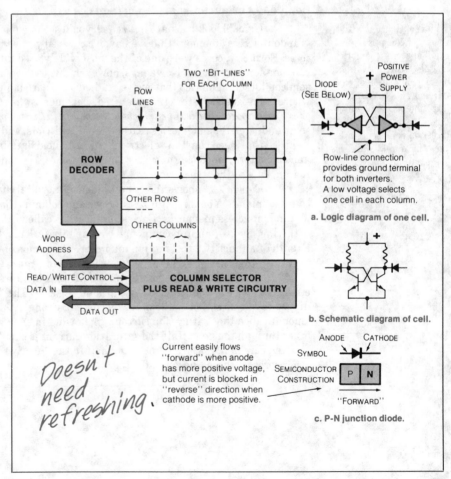

a. Logic diagram of one cell.

b. Schematic diagram of cell.

c. P-N junction diode.

To form the static cell, two inverters are cross-coupled to form a flip-flop. Information is written into or read out of the cell through the column lines.

We won't go into the electrical details of how these cells operate. What's important for you to notice is that the two inverters are *cross-coupled* like the R-S latch we studied in Chapter 4. This is what makes the circuit a flip-flop. Each inverter keeps the other in its present state, until a strong current coming in from one bit-line or the other *upsets the balance and flips the circuit to the other state*. This is how information is written into the cells, by way of the column lines. Similarly, a flip-flop is "read" via the column lines by the fact that the inverter which is in the "low" state draws current from the bit-line that's connected to its output.

A flip-flop can only be "flipped" or "read" when its row-line connection is *low* in voltage, so that's how one cell in each column is selected. The correct column, of course, is selected by the decoder circuits of the column selector, which are similar to the decoders we have learned about previously. Reading and writing are controlled for the total memory by the "read/write control" signal shown in *Figure 7-11*.

How does a magnetic-core memory work?

While constructed with rows and columns like all random-access memories, magnetic-core memories use a combination of electric currents to change the direction of the magnetic field of ferrite rings. One direction represents a 1; the other direction represents a 0.

If you'll look back at *Figure 7-2*, you'll see one more category of "random-access" memory that we haven't covered yet, which is *magnetic cores*. Formerly, cores were one of the most widely used kind of random-access memory. Nowadays, however, due to decreasing costs for semiconductor memories that we've studied, cores are mainly used only where "non-volatile" storage is required. This means that *the stored data doesn't "evaporate" (disappear) when the power is turned off*. All the semiconductor-type memories (except ROMs, PROMs, and EPROMs) are *volatile*—they need constant electrical power to hold data in their flip-flops and capacitors. Since magnetic-core memories are still widely used for this reason, let's see how they work.

Figure 7-12 shows the general idea of a typical core memory unit called a "plane." You'll recognize the rows and columns that are typical of all random-access memories. Each storage cell is called a "core." It's a little ring made of an iron compound called "ferrite," about a hundredth to a tenth of an inch (0.2 to 2 mm) in diameter. Thousands of cores are strung very close together on straight wires as you see in *Figure 7-12*.

A bit is stored in a core by *magnetizing* it in a direction pointing either clockwise or counterclockwise around the ring. The field can be flipped in one direction or the other by current stronger than a certain amount called the "saturation current" S, flowing in one direction or the other through the hole. Half the saturation current is supplied to a selected core in its "X drive line," and half in the "Y drive line," so that no other cores on these lines are affected.

Magnetic-core memories do not require any power to keep the cores properly magnetized. Thus, the memory is non-volatile, meaning the stored information is not lost when power is turned off.

A core is "read" by attempting to flip its field to the "0" direction. If the core was already in the "1" position, its magnetic field will "flip over." The changing magnetic field generates an electric pulse in the "sense wire" passing through all cores in the plane. Then the core has to be *restored* to its original "1" position, to keep the stored data the same. This involves temporarily storing the bit that was just read in a flip-flop, and routing it back to the core-drive circuitry, which immediately writes the bit back into the same core.

Since only one core at a time can be accessed in a plane, each bit of a word is stored in a different plane.

The reason core memory units are non-volatile is that it doesn't take any power to keep the cores magnetized. In some systems, it's very important for stored data to survive a power failure, without having to provide an emergency power supply. Because of this capability, cores will probably continue to be used in certain systems, even though they cost more and take up more space per bit than semiconductor memories.

**Figure 7-12.
A Plane of a Core
Memory Unit**

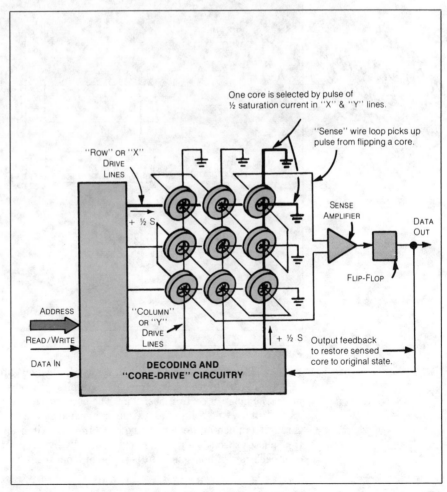

How do "peripheral" memory units work?

Speaking of cost per bit, this brings us around to the remaining four memory types at the far right end of *Figure 7-2*. These are methods that provide *huge amounts of storage capacity* needed in certain systems, at a very *low cost per bit*. But their *access times are so slow* that they're not used in the main part of a digital system. Instead, they're hooked up as *separate units* for the main system to use when necessary. They're called "peripheral" storage units, meaning "around the outside."

Figure 7-13 shows, in a highly simplified form, how these four "peripheral" memory types are similar to each other. They all store bits in *rows and columns* on a thin piece of material which *moves* past stationary devices that *read* the bits (and perhaps *write* new bits). So these methods all use *serial access* as we saw earlier for shift registers.

**Figure 7-13.
Peripheral Storage
Methods**

Large amounts of information can be stored, but access time is very slow.

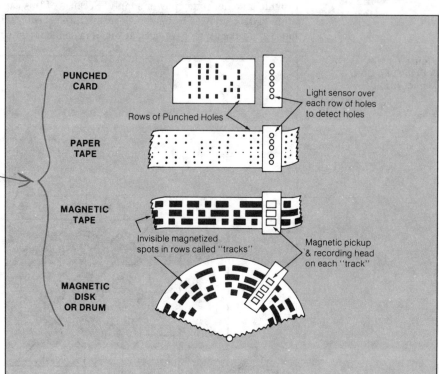

Punched cards and paper tape

There are several different kinds of punched cards and paper tape for digital systems, but they all store bits by *punching or not punching* holes in certain positions. The positions are arranged in *rows and columns*. One column at a time is punched as the card or tape moves through a punching device. And one column at a time is read as the card or tape passes through a reading device, usually by a light sensor mounted over each row of holes as light is supplied from the other side.

Information is stored in paper tape and punched cards by punching or not punching holes in a tape or card. The punched tape or card is then a form of read-only memory.

One column of bits is sometimes referred to as a "byte" (maybe because it's a bunch of bits that's "bitten off" together). "Byte" usually refers to several bits grouped together for convenience in *transmitting and receiving* them. Usually, it takes several bytes to make a *word*. As we have seen, a word is a group of bits that are *stored together in a RAM*, such as a binary number.

Since cards and paper tape are slow and comparatively difficult to handle, and since the stored data can't be *changed*, these methods are used mainly for getting information *into and out of a system*. A stack of punched cards or a reel of punched tape is read into another memory in the system, and then put back on the shelf. Occasionally, however, you'll see a *loop* of paper tape being used as a *recirculating read-only memory*.

Magnetic tape, disks, and drums

Instead of using holes or the absence of holes to store information, magnetized spots perform the same function on magnetic tape and disks. However, the tape or disks are re-useable since the magnetic spots can be erased and re-written.

As you can see in *Figure 7-13*, magnetic tape, disks, and drums contain bits stored as rows and columns of *tiny, invisible, magnetized spots*. The spots are created in a thin film of iron oxide on the surface of plastic tape or a spinning disk of aluminum or a rotating aluminum drum. (A drum is a hollow cylinder. It's not shown in *Figure 7-13* because it's so similar to a disk.) The writing and reading of bits is done in the same way sounds are recorded and played back with a tape recorder, by using a tiny electromagnet called a "head" that the surface passes under. The magnetized spots serve exactly the *same purpose as holes in a card or paper tape*, with the added advantage that the data can be *erased and re-written*.

Magnetic tape is mostly used as a rapid way to get information into and out of a system. But it's more versatile than paper tape and cards, because it stores much *more information* per square inch, it can be written and read much *faster*, and because it can be rapidly moved both *backward and forward* in the machine. Though it's written and read in serial fashion, you can start and stop at any point in the reel. However, it will typically take a minute or so to get to the starting point.

Magnetic disks and drums perform exactly the same function as *recirculating shift registers*, although much longer in *bit capacity* on each "track," and considerably *slower*. Each track is a circular row of bits, so each bit in the track passes the head once in each revolution. (Disks turn at about 1,500 revolutions per minute, and drums at up to 6,000 rpm.) Disks and drums store less information than a reel of tape, but you can get to the spot you want much quicker. These units are used as a sort of high-capacity backup or "warehouse" for information in a system.

How do cost and access time determine memory applications?

All through the chapter, we've said that the *cost per bit* and the *speed* of a particular memory unit determine how it's used in a system. Now we've learned enough that we can get more specific in our comparisons.

Figure 7-14 shows, in the form of a graphic chart, the general range of cost per bit and "access time" for several types of mass memory. (We're leaving out cards and paper tape because they're mainly input-output methods. You can ignore "charge-coupled devices and magnetic bubbles" until the last chapter.) "Access time" means the average time it takes for a bit or word to be written or read *at random* in a memory unit. (In a serial-access unit, this is half the time to go from one end of the stored data to the other.) Note that each chart division is a certain *multiple* of the neighboring divisions, so we're looking at "logarithmic" scales.

The cost per bit, the speed of access time, and the total capacity are important parameters to be considered in choosing a memory system. These vary greatly for the different types of memories.

**Figure 7-14.
Mass Memory Cost,
Speed, and Capacity
(late 1970s)**

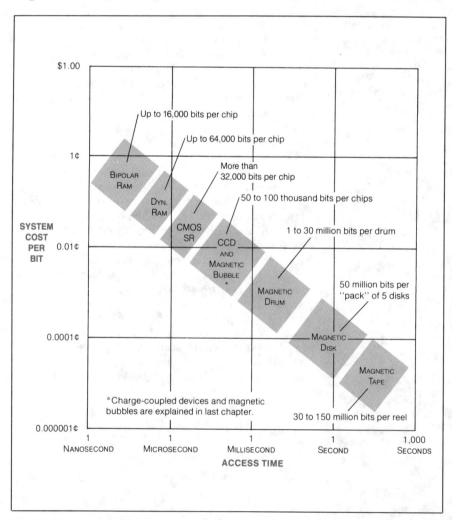

Also written in on the chart are typical *capacities* of various units, in terms of bits. Generally speaking, the lower the cost per bit, the more capacity you can afford in a given unit; so these two aspects are closely related. The main point to notice is the *very wide range* among the memory types in cost, speed, and capacity.

To give you an idea of what these numbers mean in terms of how various memory types are used, let's consider a typical computer, shown in general block-diagram form in *Figure 7-15*. For the time being, you can assume that this system works much like the calculator we've been discussing all along. (Our adder-subtracter is a sort of arithmetic and logic unit or ALU. Our number and flag registers are a sort of main memory.)

**Figure 7-15.
Mass Memory Units in a
Typical Computer**

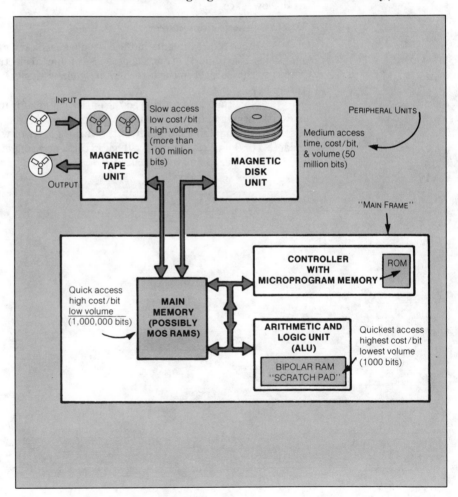

Mixing and matching the capabilities of the different memory types with the various operations required in a computer can provide the desired balance of cost, performance, size, and speed.

The point is that many systems mix and match *several different kinds of memory*, depending on the speed needed and the capacity that can be afforded. For the very quickest possible access in this computer, we can afford a little bit of bipolar RAM capacity as a sort of "scratch pad" in the CPU. For a large amount of storage that we couldn't possibly afford to put in the main memory, we use a disk unit. When we need to use data from the disk, we move a big batch of it into the main memory where we can get at it much faster. And to handle an even larger amount of data moving in and out of the system, we use magnetic tape. Thus, one of the main concerns in designing a system is to make the best use of several different kinds of mass storage unit, to achieve the desired performance at the lowest cost.

Where do we go from here?

Since improvements in the cost and speed (and capacity) of mass memories is so important for the progress of digital electronics, we'll return to this subject in the last chapter, when we look at the direction of future developments.

Now that we've surveyed the entire range of information storage in digital systems, from sequential building-blocks through mass memory units, we've completed our study of the various parts of which a system is made. So in the next chapter, we'll begin concentrating on how the parts are put together to make systems.

Quiz for Chapter 7

1. A shift register can be made into a recirculating mass memory unit by:
   a. Routing data to an adder-subtracter.
   b. Making the register a thousand bits long.
   c. Providing a data selector to load or recirculate.
   d. Using dynamic storage units.

2. A serial-access memory unit is especially useful in applications where:
   a. Data consists of numbers.
   b. Short access time is required.
   c. Each stored word is processed differently.
   d. Data naturally needs to flow in or out (or both) in serial form.

3. In a dynamic shift register, what prevents data from being lost due to leakage of charge?
   a. Preventing all leakage with good insulation.
   b. Renewing the charge with constant current.
   c. At every shift, the strength of the charge representing a bit is refreshed by the inverters.
   d. Each bit is shifted all the way to the output of the register before it has time to lose all its strength.

4. What is the principal advantage of dynamic shift registers over static shift registers?
   a. They lose their data if shifted too slowly.
   b. They can be shifted faster than any other kind of shift register.
   c. They provide more bits in smaller area.
   d. A and B above.

5. Addressing one or more RAM cells for access requires:
   a. Selecting a row line, using part of the address.
   b. Selecting one or more column lines, using part of the address.
   c. Both of the above.
   d. None of the above.

6. In an MOS read-only memory, each column-line with its transistors, neighboring ground line, and load transistor acts as:
   a. A dynamic storage cell.
   b. A static storage cell.
   c. A shift register.
   d. A logic gate.

7. In an MOS read-only memory, why doesn't the electric field from a metal row-line turn on "channels" between all the column-lines and ground lines it crosses over?
   a. The oxide is too thick except where transistors are desired.
   b. The oxide is too thin except where transistors are desired.

c. No charge is stored on the row-line.

d. The column-lines and ground-lines are too far apart.

8. "Random-access memory" or "RAM" usually refers to:

a. All memory types with random access.

b. Only those random-access-type memories that can be written into as well as read from.

c. Only memories with dynamic cells.

d. Only memories with static cells.

9. What is the main disadvantage of dynamic RAMs?

a. They can't be written into.

b. They are addressed serially.

c. You have to read the data out before it decays due to charge leakage.

d. Normal operation must include refreshing charges.

10. What feature of a static RAM cell makes it a flip-flop?

a. Its storage of charge.

b. Accessing it by row and column.

c. Two inverting gates cross-coupled as a latch.

d. The use of current for "writing" and "reading."

11. What is the principal advantage of magnetic cores over semiconductor RAM cells?

a. They are much smaller and more compact.

b. They are much more economical.

c. A bit doesn't have to be re-written after reading.

d. Stored information remains when the power is turned off.

12. Why are cards, tape, and disks considered to be serial-access rather than random-access units?

a. To get to a desired part of the stored information, you've got to pass by other parts.

b. They don't use semiconductor circuits.

c. They don't store data in rows and columns.

d. Their access time is longer.

13. How are magnetic disks and drums like recirculating shift registers?

a. They have about the same range of capacity in number of bits.

b. Their access times are about the same.

c. Their storage is non-volatile.

d. Stored data is available for reading over and over again, in the same order.

(Answers in back of the book)

How Digital Systems Function

Recall that digital information is made up of bits, and is processed by a digital system made up of gates and flip-flops.

Before we go on, let's take time for a brief review. *Figure 8-1* illustrates the high points of what we've learned about digital electronics: Digital systems process information by putting it into the form of many little pieces called "bits." These bits are processed by simple circuits called "gates" and stored in several different ways—particularly in "flip-flops" made essentially of gates. Gates and storage units can be put together in any number of different ways to handle nearly any kind of information-processing job. But there are certain general patterns that have been found useful as "building-blocks" to do certain jobs needed in a very wide variety of systems. We have learned to recognize several typical kinds of building-blocks.

**Figure 8-1.
How Do Entire Systems Work?**

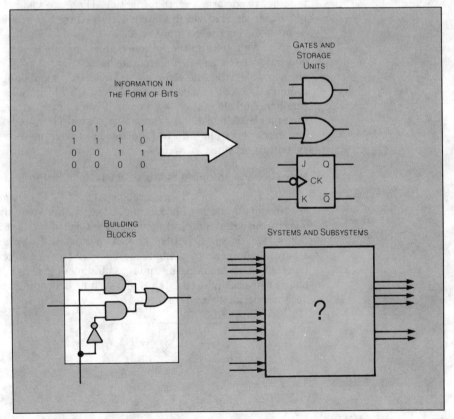

Now we're ready to move on to the next question, which is, "How do entire digital *systems* function?" That is, how do the parts of a system work together to perform the "function" (the job or purpose) of the system? How are digital systems *designed*? The same question also applies to digital *subsystems* and any other major parts of systems which have their own functions (purposes).

This is a pretty *big* question, because there are so many different *kinds* of digital systems and subsystems. They range in size and complexity from a simple digital voltmeter as in *Figure 4-2*, to computers. (*Figure 7-15* is one example, but many computers are much more complex.) Information may flow in and out very rapidly as in the digital television transmission system that we envisioned in *Figure 5-15*, or as slowly as your fingers press the keys of a calculator. Some systems such as the autopilot of *Figure 5-11* do only one job, while others such as computers can do many different jobs. So how can we hope to comprehend so many different things?

How can we simplify our study of digital systems?

Since we can't hope to turn you into a digital system designer in this chapter, the approach that we'll take is to show you *three important factors* that a system designer has in mind when he puts building-blocks together to perform a certain function. These are some of the most important alternatives in system design—three basic decisions that must be made during the design process.

We'll take up these three factors one at a time, in the order in which they're easiest to understand, based on what you have already learned. In illustrating these points, we'll see a number of *examples* of combinations of various building-blocks with which we have become familiar. And along the way, we will become much better acquainted with the inner workings of our example calculator. Thus, we will build a general understanding of how digital functions are performed—how digital systems function.

How does the grouping of bits affect system design?

One of the first questions when choosing a digital design is, "What size group of bits should the system handle?"

First, there's the general question of *how many bits we work on at one time* in any stage of processing. Of course, we know that information goes in and out of a digital system in the form of *groups* of bits. But inside the system, the groups can be broken up into *smaller groups* processed separately (or perhaps combined into *larger groups* and processed together). This breaking up and recombining of groups of bits may take place at several different stages in a system, depending on what the designer decides works best.

Loosely speaking, we're talking about the choice between *parallel* and *serial* processing for the same function. To show the alternatives involved, let's take a familiar example function of binary addition, which we first discussed in Chapter 3. First, let's look at *full parallel* addition—meaning that all the bits involved are added at once, without being broken into smaller groups.

What's involved in "full parallel" binary addition?

When adding two 16-bit numbers with a full parallel binary adder, all bits are added in one step.

Figure 8-2 shows what is involved when two 16-bit numbers are added in "full parallel" fashion. This approach would be employed in the "arithmetic and logic unit" or ALU of a computer that uses 16-bit words. (ALU is pronounced by spelling A-L-U. An ALU is a sort of extra-sophisticated adder-subtracter, as we will see in the next chapter.) Two words to be added are routed from the random-access main memory (not shown), placed in two parallel 16-bit "input registers," and then routed to the 16-bit adder. The resulting 16-bit sum word goes into an "output register," from which it is later routed to be stored in the main memory.

**Figure 8-2.
An ALU 16-bit Adder**

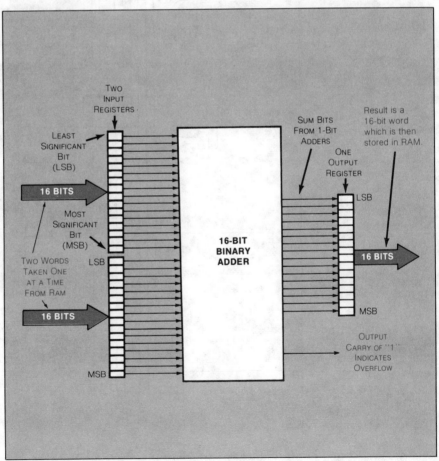

**Figure 8-3.
A Carry-Look-Ahead
Binary Adder**

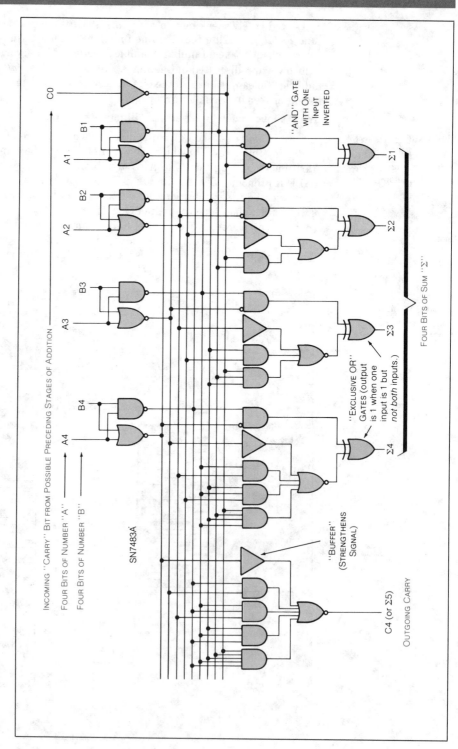

INCOMING "CARRY" BIT FROM POSSIBLE PRECEDING STAGES OF ADDITION

FOUR BITS OF NUMBER "A"

FOUR BITS OF NUMBER "B"

SN7483A

"AND" GATE
WITH ONE
INPUT
INVERTED

FOUR BITS OF SUM "Σ"

"EXCLUSIVE OR"
GATES (output
is 1 when one
input is 1 but
not both inputs.)

"BUFFER"
(STRENGTHENS
SIGNAL)

C4 (or Σ5)

OUTGOING CARRY

Now the point of this example is to show that *all the bits* of two binary number can be added in just *one step*. The groups of bits involved— the words, representing binary numbers—are *not* broken up into smaller groups as we mentioned a little bit earlier.

A special circuit called carry-look-ahead is used in binary adders to allow the adder to react faster to carries.

In a moment, we'll see what's involved when we *do* break up numbers for addition. But first, you may be curious as to how the adder in a computer might work. In a computer's ALU, we want to add as quickly as possible. So the ALU would probably use a very fast type of design called a "carry-look-ahead" adder. Even though a 16-bit carry-look-ahead adder is too complicated for us to look at, it would work something like the *four-bit* carry-look-ahead adder shown in *Figure 8-3*. It would be much more complicated and would handle numbers with 16 bits rather than four. Study the adder in *Figure 8-3* to see that it does exactly the same job as the one we saw in *Figure 3-14*, but it completes an addition *faster* because it has special circuitry to handle carries so it doesn't have to wait for "carry" bits to pass (ripple) between one-bit adders.

To get back to our subject now, let's see what's involved in adding numbers in *full serial* fashion.

What's involved in "full serial" binary addition?

"Full serial" addition of binary numbers means adding *one bit at a time* from each of the incoming numbers, using the *same one-bit full adder* for all the bits. (Refer to Chapter 3 for a refresher on one-bit adders.) *Figure 8-4* shows how this may be done in the case of adding two 16-bit numbers.

**Figure 8-4.
Full-Serial Addition of
Binary Numbers**

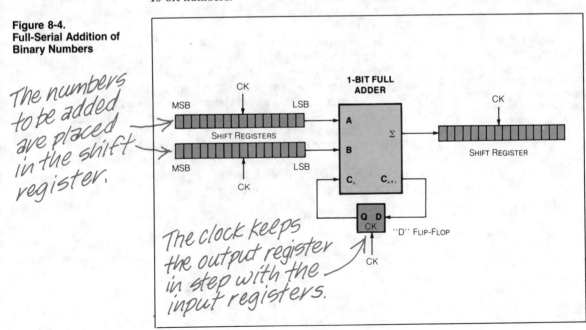

The numbers to be added are placed in the shift register.

The clock keeps the output register in step with the input registers.

Full serial binary addition only uses a 1-bit adder and three shift registers to perform the addition function one bit at a time.

The two incoming numbers to be added, called A and B, are first placed in two shift registers (by way of connections not shown in *Figure 8-4*). The two registers are clocked in step with a third shift register that receives the "sum" bit. The least-significant bits (LSBs) go to the adder first, and after 16 clock pulses, the 16-bit sum is in the output register. A flip-flop clocked in step with the registers stores the "carry" bit from adding each pair of bits, and passes it to the adder for adding in with the next pair.

(Incidentally, although the adders in this chapter perform the same function as those in Chapter 3, you'll see slightly different labels for the inputs and outputs in *Figures 8-3* and *8-4*. These somewhat more complicated labels match those you're likely to see in an integrated-circuit catalog.)

Full-serial addition wouldn't be used in the arithmetic and logic unit of a computer, because it's far too slow. But it comes in very handy when we don't need high speed but *do* need to save cost and space on an IC chip. A good example is the "program counter" in our example calculator—so let's take a look.

How would a program counter use serial addition?

In any system that operates by following a series of stored instructions (such as our calculator), the "program counter" consists of the address register, together with circuitry to *add one to the address in the register* when required. This is so that the next instruction in a series can be called up for the next instruction cycle, enabling the controller to step through a programmed "routine" as we discussed in Chapter 1. *Figure 8-5* shows how the program counter in our calculator could use full-serial addition for this purpose.

The principles of operation of the program counter and the full-serial adder are basically the same, except that the address register is an input register and a sum register all in one.

If you compare the program counter with the full-serial adder in the preceding figure, you'll see that the principle is the same. However, in the program counter, the *address register* serves both to provide one input number "B" and to receive the sum. (This is an 8-bit shift register with parallel inputs and outputs as we discussed with respect to *Figure 4-3b* and *4-6*. We assume our calculator's addresses consist of 8 bits each.) The other input number "A" is provided by the "addend" flip-flop. This unit has a constant "0" at its data input. But it can be preset to "1" by the controller at the beginning of each instruction cycle. (You'll remember from Chapter 1 that an instruction cycle is the time during which the controller uses each instruction it calls up.)

**Figure 8-5.
A Full-Serial Program
Counter**

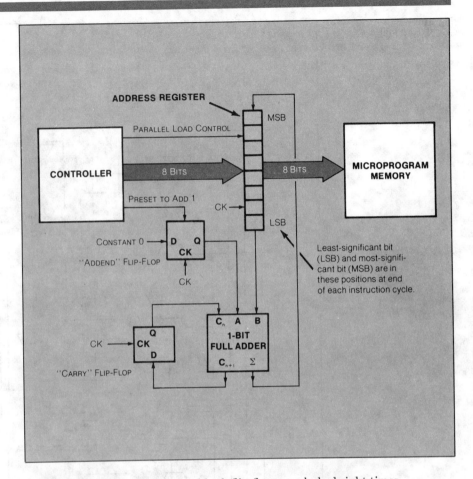

The address register and both flip-flops are clocked eight times
during each instruction cycle. If nothing is to be added to the current
address (say, to repeat the current instruction), the controller *does not
preset* the addend flip-flop. So the constant 0 is added to each bit of the
address as it recirculates through the adder. At the end of the instruction
cycle, the address is back in the correct position for a short time, during
which the microprogram memory refers to it and fetches the instruction
at that address.

To add 1 to the address, the controller presets the addend flip-flop
to 1 at the beginning of an instruction cycle. This 1 is added to the LSB of
the address during the first clock cycle. For the remaining seven cycles,
the addend flip-flop presents the constant 0 to be added to the other seven
bits of the old address.

The controller pre-sets a 1
into a D flip-flop when a 1
is to be added to the ad-
dress register. Otherwise
a 0 is added.

**Figure 8-6.
A TTL 1-bit Full Adder**

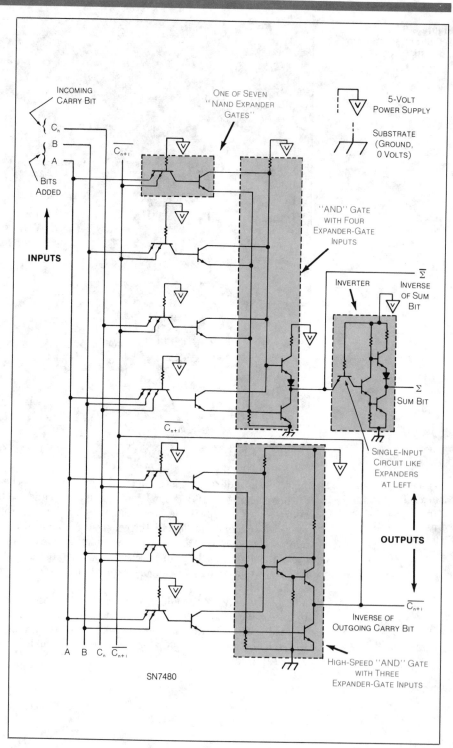

SN7480

How complex is a one-bit adder?

The combinational logic
circuitry of a 1-bit adder is
made up of AND-OR-IN-
VERT functions.

The difference between the complexity and size of a 16-bit adder
and a 1-bit adder may not seem very important unless you remember how
building-blocks like these are made. To make sure we don't loose sight of
the inner workings of the parts we're discussing, let's look at a schematic
diagram of a one-bit adder, shown in *Figure 8-6*.

This is part of the circuitry of a Texas Instrument's IC called
SN7480, taken directly from *The TTL Data Book* with minor
rearrangement for easier understanding. (The calculator uses MOS
circuitry instead of TTL, but the example is still meaningful.) As an
exercise, see if you can draw the logic diagram for this circuit by
inspecting *Figure 8-6*. (It's not fair to peek at *Figure 8-7* until after you're
finished!)

In picking the gates out of this TTL circuit, you'll recognize seven
"NAND expander gates" in two groups, with each group arranged much
as we saw in *Figure 6-9*. Beside having a common output resistor as in
Figure 6-9, each group of NAND expanders in *Figure 8-6* is followed by
circuitry shown in the two middle outlined boxes. This circuitry improves
noise margin and fan-out (the ability to drive other circuits without
making errors due to "noise.")

Each group of expander gates with the associated circuitry to the
right acts as a group of NAND gates feeding an AND gate. As we learned
back in *Figure 6-9*, this combination acts as a group of AND gates feeding
a NOR gate, which is called an "AND-OR-invert" combination. As for the
inverter in the outlined box at the right in *Figure 8-6*, note that it consists
of one single-input TTL expander circuit followed by circuitry like that in
the upper middle outlined box.

What does the logic diagram look like for the adder?

Note the complexity of the
1-bit adder. This would be
reproduced for each bit if
the digital system oper-
ated using parallel data.

For your reference, *Figure 8-7* shows the logic diagram for the
schematic in *Figure 8-6* (assuming positive logic). Also shown is the truth
table, which happens to be the *same whether positive or negative logic is
used*. The basic part of the truth table is the same as we first saw back in
Figure 3-15. But the logic diagram in *Figure 8-7* is different from the one
in *Figure 3-15*. This is to take maximum advantage of the simple, fast
"AND-OR-invert" combination, as explained in *Figure 8-7*.

The main thing you should note in *Figure 8-7* is that *each gate
can be recognized in the schematic diagram*. Beside improving your
familiarity with TTL schematic diagrams, this example is a reminder of
how complex even a simple one-bit adder can be. So you can see why we
want to avoid full-parallel addition if we can get away with it.

Figure 8-7.
Logic Diagram and Truth
Table for 1-bit Adder

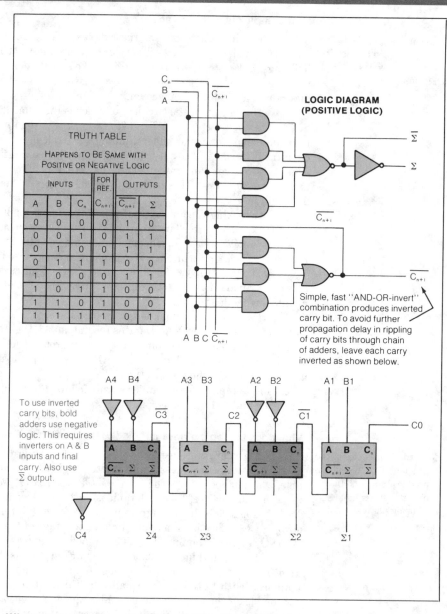

TRUTH TABLE					
HAPPENS TO BE SAME WITH POSITIVE OR NEGATIVE LOGIC					
INPUTS			FOR REF.	OUTPUTS	
A	B	C_n	C_{n+1}	$\overline{C_{n+1}}$	Σ
0	0	0	0	1	0
0	0	1	0	1	1
0	1	0	0	1	1
0	1	1	1	0	0
1	0	0	0	1	1
1	0	1	1	0	0
1	1	0	1	0	0
1	1	1	1	0	1

LOGIC DIAGRAM
(POSITIVE LOGIC)

Simple, fast "AND-OR-invert" combination produces inverted carry bit. To avoid further propagation delay in rippling of carry bits through chain of adders, leave each carry inverted as shown below.

To use inverted carry bits, bold adders use negative logic. This requires inverters on A & B inputs and final carry. Also use $\overline{\Sigma}$ output.

Bit-parallel, digit-serial bit processing is a method of binary addition that is between full bit-parallel and full digit-serial processing. BCD numbers are very effectively processed in this manner.

What are the choices other than full parallel or serial?

In many cases, the system designer will find it best to use a grouping of bits *in between* the extremes of full-parallel and full-serial. A good example of this is the adder-subtracter in our calculator, which we last discussed with regard to *Figure 7-5*. A simplified picture is provided in *Figure 8-8*. This "combination" method of grouping bits in a BCD number is called "bit-parallel, digit-serial."

**Figure 8-8.
Bit-Parallel, Digit-Serial
Addition**

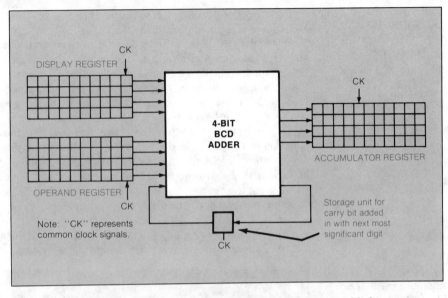

*This is a way
of processing
information
that is not
full parallel
or serial.*

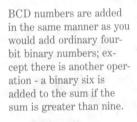

As we have seen before, one pair of four-bit groups is added at a time, representing a pair of decimal digits. The "carry" bit is stored between addition steps in a flip-flop clocked in step with the shift registers that handle the numbers. Although *binary* numbers can be handled in groups like this, such an approach is particularly convenient for *binary-coded decimal* numbers, which the calculator uses.

How do you add BCD numbers?

BCD numbers are added in the same manner as you would add ordinary four-bit binary numbers; except there is another operation - a binary six is added to the sum if the sum is greater than nine.

We haven't yet seen how BCD numbers are added, so let's take a quick look at how it's done. It's just a matter of adding digits as if they were ordinary four-bit binary numbers, and then *adding six if the sum is greater than nine.*

Figure 8-9 shows an example of how this works, in the addition of nine and eight. The result of binary addition is 10001, which is binary seventeen. Since this is greater than nine, we add six, giving us binary twenty-three, which is 10111. This result is treated as a carry of 1 and a BCD sum of 0111, or seventeen in BCD form. Simple, isn't it?

**Figure 8-9.
Correcting a Binary
Number to BCD Form**

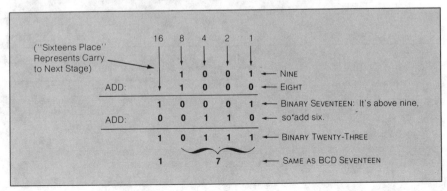

Two four-bit binary adders are used to add BCD numbers in the calculator. The first adder produces a sum and a decoder detects if the sum is greater than nine. If it is, a binary six is added to the sum in the second adder.

Figure 8-10 shows how this method for BCD addition can possibly be handled in a calculator. Inside the adder-subtracter would be *two* 4-bit binary adders as shown. A little decoder network puts out a "1" when the sum is greater than nine, and a "0" otherwise. In the second adder, the binary sum is added to a number consisting of "0" in the ones and eights places, and the *decoder output in the twos and fours places*. So this number is six (0110) when the decoder is recognizing a number greater than nine. Again, it's a pretty simple idea, right?

What you've seen in these discussions of how the addition function can be performed is examples of different ways to *group bits together* for processing. The question of grouping comes up in the design of various stages of processing in a system—not just for addition. And as we've seen, the answer depends on careful consideration of *how fast* the function must be performed and *how much processing capacity can be afforded*.

So now we're ready to move on to the second of the three factors that we said must be considered in designing a "system function."

Figure 8-10.
Additions in BCD Code
in a Calculator

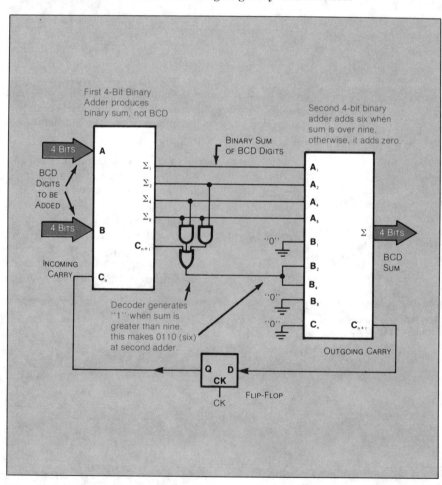

In a hard-wired system,
the way the circuit oper-
ates is permanently fixed.
In a variable-programmed
system, the operation can
be changed by changing
the program.

What are "hard-wired" and "variable-programmed" control methods?

The second factor or alternative is whether to perform a given sequence of functions by using "fixed" or "hard-wired" control on the one hand, or "variable programming" on the other hand. A hard-wired system (or subsystem) is one whose function or behavior is *permanently fixed,* so that the way it operates and the things it does *cannot be changed.* The function is "wired" into the system, by the way the "hardware" (the transistors, gates, components, wires) is connected together. But a "variable-programmed" system (or subsystem) *follows instructions from a memory unit* in deciding what to do. So its operation can be *changed* by changing the "programs" in the memory. Such programs are called "software," because they're different from the "hardware" involved in hard-wired systems.

The difference is not that hard-wired systems don't have programs. Many of them *do* follow steps in a sequence, and these steps are often called a "program" The difference is that the program *cannot be changed* in a hard-wired system.

To give you an idea of the difference between "hard-wired" functions and "variable-programmed" functions, let's look at an example of each kind in our calculator.

What does the hard-wired segment decoder subsystem do?

One important and fairly complicated hard-wired function in the calculator is that performed by the *segment decoder* subsystem. This function is illustrated in *Figure 8-11.* First, let's see what the segment decoder subsystem *does,* and then we'll see how it does this by using *hard-wired* programming.

As indicated in *Figure 8-11,* the four output bits from the recirculating display register are supplied to the decoder. (We discussed recirculating registers in *Figure 7-5.* The routing details will be covered later in this chapter.) To understand the function of the decoder, we must have in mind exactly how numbers are stored in the register, as shown in *Figure 8-11.*

Because of its continuous recirculation, the display register has ten storage spots for four-bit BCD digits, which revolve around and around like the baskets on a Ferris wheel. So we'll call these spots "digit baskets," and give each basket a number as shown in *Figure 8-11.* Baskets 0 through 7 are for the actual digits of the number, not counting the sign or decimal point. These eight digits together are called the "integer" for the number. (An integer is a number without a decimal point.) Basket 0 is for the least-significant digit (LSD), and basket 7 is for the most-significant digit (MSD).

Basket 8, in turn, is where we would put a "1" to store a "minus" (negative) sign for the number. And basket 9 is for the "decimal-point digit" or DPD. The DPD tells *how many steps from the right end of the integer* the decimal point goes.

Above the display in *Figure 8-11*, we see the basket number for the digit that's displayed in each of the nine character positions. (Note that the character position at the far left is for a minus sign or symbols indicating errors, overflow, etc.) The decimal-point LED (light-emitting diode) for each character position is at the *right* of the digit. So the DPD is the basket number for the digit that's displayed in the character position where the decimal point goes. As an example, notice that a DPD of 3 and an integer of 00007426 in the display register cause "7.426" to be shown in the display.

The hard-wired segment decoder receives digits from the display register one digit at a time and displays the digit during the time the scan line is on.

**Figure 8-11.
A Calculator's Segment
Decoder Subsystem**

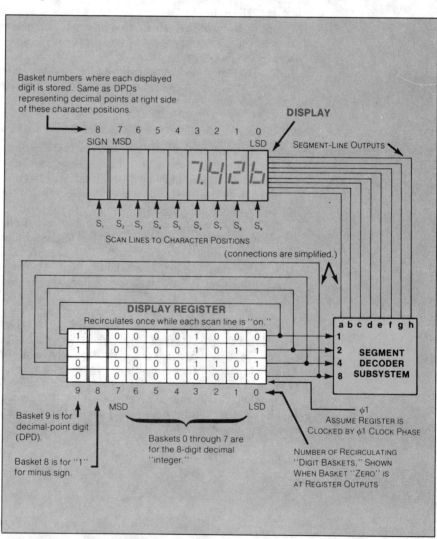

You'll remember from *Figure 1-5* that one character position at a time is illuminated when its scan line is "on." The scan lines (labelled S_1 through S_9) are turned on one after another, from left to right. Every time a new scan line is turned on, the decoder subsystem must switch its eight outputs to the pattern for the digit (and maybe decimal point) to be displayed in this next character position.

The basket (digits) of the display register recirculate within the register when each scan line is on. After all digits are scanned, the display register may be changed.

Meantime, the display register makes *one full recirculation* while each scan line is "on." This constitutes one instruction cycle, the time during which one instruction is in effect. So during each instruction cycle, the decoder subsystem has to *pick out and store* the integer digit that's to be displayed next. It also has to look at the DPD, decide whether it's going to turn on the decimal point for the next character position, and remember this decision. The digit and decimal-point signals that it's currently transmitting are the ones it stored this way during the *preceding* instruction cycle.

That's a summary of *most* of the things the segment decoder subsystem does. It has still other jobs to do, such as providing a minus sign or error characters, and leaving out zeroes at the left of a number. But just the simplified function of getting the correct digits and decimal point into the correct character positions is enough to provide an illustration of the idea of "hard-wired" control.

What's a possible design for the segment decoder subsystem?

Figure 8-12 shows a possible design for a subsystem to perform this simplified function. As you can see, even a "simple" hard-wired function can get pretty complicated. In the upper left-hand corner, you'll see the 7-segment decoder-driver that we studied in Chapter 3. Four-bit latch "D" holds the digit currently being decoded and transmitted, and four-bit latch "C" stores the digit picked out for transmitting during the *next* instruction cycle. Similarly, flip-flop "F" transmits a "1" during the instruction cycle when the decimal-point LED is illuminated. And flip-flop "E" stores a "1" when the *next* instruction cycle will have the decimal point turned on. (The "decimal-point driver" provides segment output "h" by accepting or "sinking" current from the LED to turn it on.)

The remaining parts of the subsystem (in the large block) provide the "control signals" (shown as bold lines) for the main working parts that we just named. Together with *timing signals* (the three clock phases), these signals are what make the right things happen at the right times in the subsystem. The point that we're illustrating by learning about this subsystem is that *these control signals are generated in a "hard-wired" manner*, rather than as a result of stored instructions.

The example shows that even fairly simple hard-wired functions can get complicated very quickly.

Enough labels and comments are provided in *Figure 8-12* for you to pick your way through the action of this subsystem if you like. You will find that the sequence of control signals is based on a series of two *counters* and a clock-signal input. This is because something different has to happen whenever a new "basket" appears at the output of the display register, and whenever a new scan line is turned on at the beginning of an instruction cycle. In effect, the control circuitry *counts baskets and scan lines*. Then it uses a decoder and some other combinational networks to figure out what to transmit in the four control lines at each new step in the count. As we will see later, this general pattern of counters followed by decoders and other combinational networks is typical of most hard-wired systems and subsystems.

How do control signals differ from timing signals?

Control signals say what to do in a system, and the timing signals say when to do it.

Figure 8-12 provides an illustration of the difference between "control" signals and "timing" signals, with regard to the operation of latches C and D. Like most systems of any complexity at all, our calculator is a "synchronous" system. This means that nearly everything that takes place is made to wait for *just the right moment in time*, as counted out by the system's three-phase clock. This is to make sure that everything happens in the proper sequence—that a new step is not begun before the preceding step is completed. In effect, *control* signals say *what* to do, and *timing* signals say precisely *when* to do it.

Notice in *Figure 8-12* that the clock signal to each latch is the AND function of a control signal and a timing signal. As the timing diagram in *Figure 8-12* indicates, the beginning of each control pulse is rather *indefinite*—due, let's say, to rippling in the counters (which we discussed in Chapter 4). This could be a problem at the beginning of "scan time 8," as scan line number eight comes on, because both latches are to be clocked during "basket zero" time, but latch D must be clocked *first*.

Making the clocking of the latches dependent on the clock phases as well as the control signals takes care of this problem, making sure that latch D gets clocked first in this "close call" situation. In effect, a control signal says, "Clock the latch," and a timing signal comes along then and says, "Okay, do it *now!*"

And that completes our example of a "hard-wired" subsystem. We've seen different things made to happen at different time in a prearranged sequence, based on counting clock pulses and decoding the counts. And we've seen the partnership arrangement between control and timing signals in a synchronous system.

Figure 8-12.
An Example Segment
Decoder Subsystem

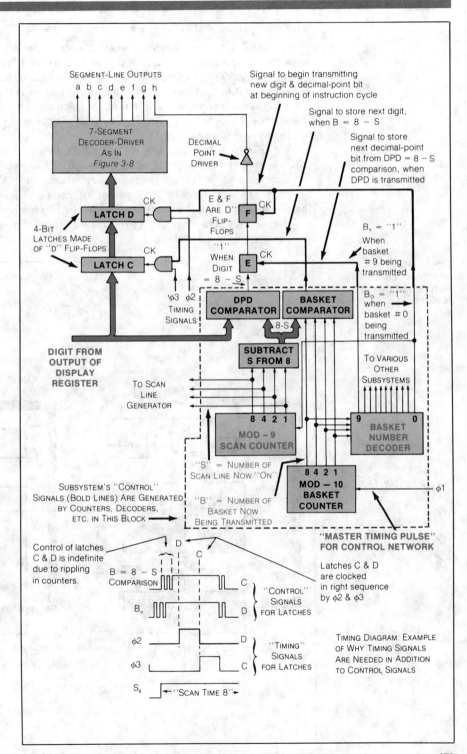

Signal to begin transmitting
new digit & decimal-point bit
at beginning of instruction cycle

Signal to store next digit,
when B = 8 − S

Signal to store
next decimal-point
bit from DPD = 8 − S
comparison, when
DPD is transmitted

SEGMENT-LINE OUTPUTS
a b c d e f g h

7-SEGMENT
DECODER-DRIVER
AS IN
Figure 3-8

DECIMAL
POINT
DRIVER

E & F
ARE D''
FLIP-
FLOPS

F

CK

CK

4-BIT
LATCHES MADE
OF "D" FLIP-FLOPS

LATCH D

LATCH C

CK

CK

"1"
WHEN
DIGIT
= 8 − S

E

CK

B_9 = "1"
When
basket
9 being
transmitted

φ3 φ2
TIMING
SIGNALS

DPD
COMPARATOR

BASKET
COMPARATOR

8-S

B_0 = "1"
when
basket # 0
being
transmitted

DIGIT FROM
OUTPUT OF
DISPLAY
REGISTER

SUBTRACT
S FROM 8

TO VARIOUS
OTHER
SUBSYSTEMS

TO SCAN
LINE
GENERATOR

8 4 2 1
MOD − 9
SCAN COUNTER

9
BASKET
NUMBER
DECODER

0

"S" = NUMBER OF
SCAN LINE NOW "ON"

8 4 2 1
MOD − 10
BASKET
COUNTER

φ1

SUBSYSTEM'S "CONTROL"
SIGNALS (BOLD LINES) ARE GENERATED
BY COUNTERS, DECODERS,
ETC. IN THIS BLOCK

"B" = NUMBER OF
BASKET NOW
BEING TRANSMITTED

"MASTER TIMING PULSE"
FOR CONTROL NETWORK

Control of latches
C & D is indefinite
due to rippling
in counters.

D

C

B = 8 − S
COMPARISON

C

Latches C & D
are clocked
in right sequence
by φ2 & φ3

B_0

D

"CONTROL"
SIGNALS
FOR LATCHES

φ2

D

"TIMING"
SIGNALS
FOR LATCHES

TIMING DIAGRAM: EXAMPLE
OF WHY TIMING SIGNALS
ARE NEEDED IN ADDITION
TO CONTROL SIGNALS

φ3

C

S_8

"SCAN TIME 8"

What's an example of "variable-program" control?

And this brings us to the other alternative for controlling sequences of operations—which we said earlier is called "variable programming." This method will be easy to understand, because we've already seen it in operation in the calculator. The fact is that the "master control" over all the various sections of the calculator (which are themselves "hard-wired"), is exerted by following *instructions stored in the microprogram memory*.

Figure 8-13 shows a highly simplified, general comparison of variable-program and sequential hard-wired control. On the right, we see the basic concept we studied in the decoder subsystem—a counter followed by decoder networks, possibly with other information also supplied. Note that in some systems, the counter can be *preset* to a beginning state that may be required. The step-by-step control sequence is "hard-wired" into the circuitry of such a system.

**Figure 8-13.
Variable-Program vs
Hard-Wired Control
Methods**

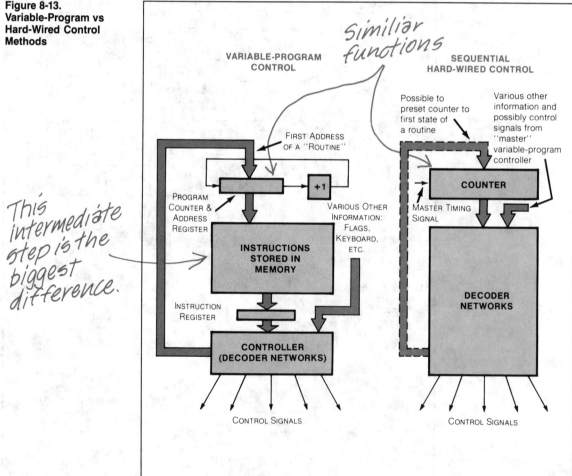

Hard-wired and variable programmed systems operate similarly. The big difference is that variable programmed systems follow instructions stored in memory rather then stepping through controller conditions that are hard-wired. The instructions can be changed easily.

The picture on the left is just a rearranged version of some parts from the calculator block diagram *(Figures 7-1 or 1-6)*. The controller loads the first address of an instruction sequence (a "routine") into the address register. The program counter steps through the routine like the hard-wired counter at the right, until the controller interferes by stopping the counting or loading a new address. Each address eventually gives rise to a certain set of control signals, as does each counter state in the hard-wired system. The controller consists mainly of decoder and other combinational networks similar in concept to those in the hard-wired system. The big difference is that the control process in the programmed system passes through the intermediate state consisting of *stored instructions*. As we will see in the next chapter, this makes all the difference in the world.

The only serious drawback of variable-program control is that it's generally *slower*, due mainly to the time required to fetch instructions from a memory. In many cases, speed requirements *prevent the use* of variable-program methods.

What examples illustrate "dedicated" and "multipurpose" functions?

Now let's move on to the last of the three system-design factors we set out to identify, which is the choice between using "dedicated" hardware and "multipurpose" hardware for performing a certain function. That is, do we use a special unit off to itself, dedicated to just one job—or do we use a unit that can do this job and others as well?

Systems and subsystems can be designed to be either dedicated to one function, such as the adder in the program counter, or multipurpose and able to perform several different tasks, such as the adder-subtractor.

Here again, we can find excellent examples of the two choices in the calculator, with respect to the *addition* function. On the one hand, the *program counter (Figure 8-5)* contains a *dedicated* adder. It's limited to a single, very special job—that of adding one to the address register ("program counter") so as to go to the next instruction. On the other hand, the *adder-subtracter* is a *multipurpose* unit that can perform the addition function. It's multipurpose not only with regard to addition, subtraction, and comparison, but also with regard to where the added numbers come from and where the sums go. We could even make a calculator in which the *adder-subtracter* is used to add one to the present address, thereby using a multipurpose adder rather than a dedicated one.

However, that would make the programming very clumsy, clutter up the microprogram memory with extra instructions for adding to addresses, and greatly increase the time required for the calculator to do anything. So as you can see, the decision between dedicated and multipurpose depends on the circumstances.

Simple functions needing little hardware usually are better served by dedicated systems. Large complex functions usually need multipurpose systems for maximum efficiency and lower cost.

Ordinarily, a very simple function requiring very little hardware is best performed in *dedicated* fashion. For example, the adder in the program counter is a relatively insignificant part of the calculator chip. But if you need a sizeable piece of hardware for the function, you'll get more for your money if you *share it* with other applications in the system. This is why we don't stick in *separate units* for adding, subtracting, and comparing. With just a little bit of extra hardware and programming (which we call "software," remember?), we can make a subsystem that will do *all three* jobs, which we call the "adder-subtracter."

Can the calculator provide further examples of system design?

That completes the general picture of the three main design factors which need to be considered in making the parts of digital systems work together. We've covered the choices between parallel and serial processing, between hard-wired and variable-programmed control, and between dedicated and multipurpose functions. You probably can see more clearly now why we've called this chapter, "How Digital Systems Function." And you can see the sort of things that are involved in putting together various building-blocks to perform certain "functions," or jobs.

To help fix the three design factors in your mind, and to provide further examples, let's look deeper into the design and operation of our example calculator. For the first time now, we're in a position to see how the calculator works in rather fine detail, and to get more specific about a number of points that we've only glossed over earlier in the book.

How do you subtract BCD numbers using "ones complement"?

Basically BCD subtraction is really BCD addition (with some adjustments) with the sign of one of the numbers changed by inverting it (taking its complement). The minuend is the number that is subtracted from, and the subtrahend is the number subtracted.

First, let's see how the adder-subtracter could manage to *subtract* binary-coded decimal numbers. *Figure 8-14* shows the principle of one of several methods that might be used. The method is called "ones-complement" subtraction because it involves using the "ones complement" of the subtrahend. (The "subtrahend" is the number *subtracted*. The "minuend" is the number *subtracted from*.) To make a number into its ones-complement, you simply *invert all the bits*—change ones to zeroes and zeroes to ones. For purposes of addition, this has the effect of *changing the sign* of the number. So we can subtract a number by adding its ones-complement, in a certain way.

As you can see in the logic diagram of *Figure 8-14*, the four-bit subtrahend digit and the incoming carry (now called "borrow") are first *inverted*. The resulting bits are *added to the minuend digit* by a binary adder. The resulting "intermediate sum" digit goes to a second adder. There, it's added to 1010 (ten) *if the first addition produced a carry (borrow) of zero*. If this first (or "intermediate") carry (borrow) bit was 1, the second adder adds nothing (0000).

**Figure 8-14.
Bit-Parallel, Digit-Serial,
Ones-Complement BCD
Subtracter**

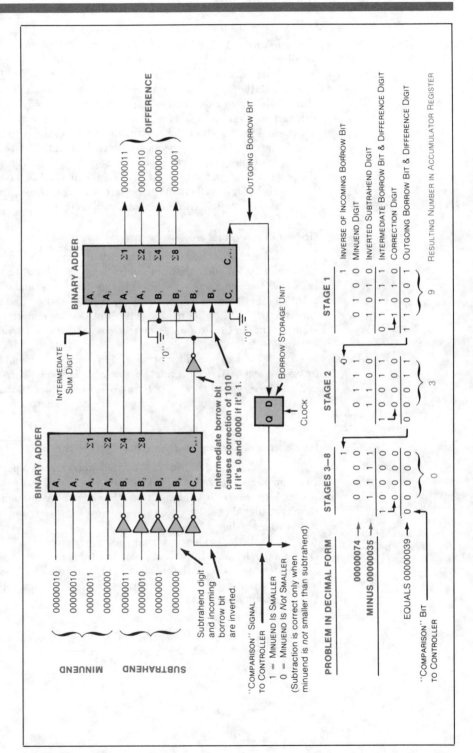

Believe it or not, this weird arrangement really subtracts. It even produces an "outgoing borrow" bit of 1 when the minuend digit is smaller than the subtrahend digit. This borrow bit is held over in the "carry-borrow" storage unit, like the "carry" bit during addition.

Rather than explaining *why* you get subtraction when you invert and add, let's just look at an example, also shown in *Figure 8-14*. We're subtracting 35 from 74. You see the various bits in the order in which they are clocked in and out of the subtracter above. At the bottom of the figure, the details of the process are listed. There, the first and second additions of each stage are listed one above the other. You should follow the process through for yourself, from right to left. We have eight stages, for the eight digit baskets in the registers. Stages three through eight are identical. The original incoming borrow bit is 0.

Does this subtracter work in every possible case?

As it turns out, this subtracter works right only when the minuend is *not smaller* than the subtrahend—that is, when the minuened is *greater than or equal to* the subtrahend. To understand why, note that after all eight pairs of example digits are shifted through the substractors, the last "borrow" bit left in the storage unit is 0. This signifies that in this example, the *whole minuend is not smaller than the subtrahend.*

If we had instead subtracted 74 from 35, we would have gotten an answer of 99999961, with a final borrow of 1. This final borrow would tell us that the *minuend is smaller than the subtrahend,* and that we're going to have to take some extra steps to get the right answer. So how do we do that?

How can "comparing" numbers help get the right answer?

Adjustments must be made if the minuend is larger than the subtrahend. When a borrow occurs in the last digit position, the minuend is made the subtrahend and vice versa, and the subtraction is done again.

As indicated in *Figure 8-14*, the subtracter also performs the function of a *comparator*, which we mentioned earlier. The "borrow" bit is *routed to the controller*, telling it whether or not the minuend is smaller than the subtrahend. This is our clue to getting the right answer when the minuend is smaller.

If the controller finds that the final borrow is "1" after first trying a particular subtraction, its instructions tell it to *try the subtraction again*, with the two incoming numbers *crossed over the other way* by the routing circuitry. This time, the final borrow will be "0," and the result will be correct.

**Figure 8-15.
Complete Adder-
Subtracter Subsystem**

*Addition, subtraction and
comparison can be preformed
with just 4-bit adders*

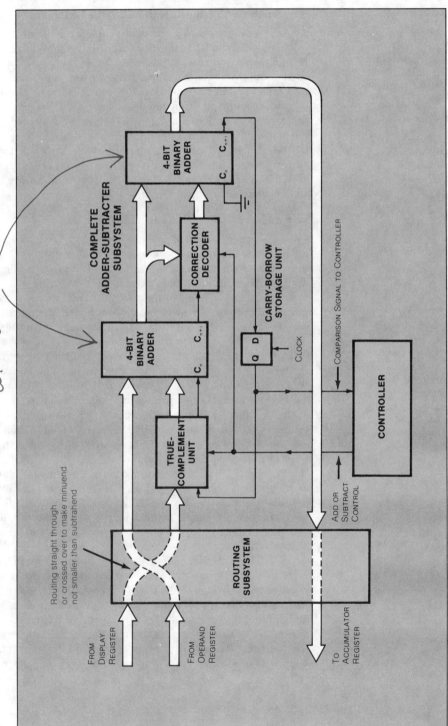

How are three functions combined in the adder-subtracter?

The adder-subtractor is a multipurpose subsystem that performs addition, subtraction and comparison functions.

As *Figure 8-15* shows, the complete adder-subtracter subsystem would *combine* the BCD addition, subtraction, and comparison functions from *Figures 8-10* and *8-14*, using the *same two 4-bit binary adders*. An "add-or-subtract" control signal from the controller causes the "true-complement" unit to either *pass through* the incoming digit and carry-borrow bit for an addition operation, or *invert* these signals for a subtraction operation. Similarly, the "correction decoder" generates either the 0110 or 0000 required for BCD addition or the 1010 or 0000 required for BCD subtraction. This, then, is how one multipurpose subsystem could perform any of three different functions—digit-serial BCD *addition*, digit-serial BCD *subtraction*, or *comparison* of digits or whole numbers—all under variable-program control.

How does a calculator handle negative numbers?

Our discussion of subtraction may make you wonder how a calculator handles *negative numbers*, meaning numbers less than zero (numbers with a minus sign). We've said from time to time that one of the digit positions in each number register is for storing a "1" to indicate the number is negative (basket number eight in *Figure 8-11*). But when the calculator performs an arithmetic operation—add, subtract, multiply, or divide—how does it know what sign to put on the resulting number?

Instructions stored in memory lead the controller through a check of the sign of the numbers it is dealing with, which number is larger, and some simple logic, to determine the sign of the result.

The answer is that the controller simply follows a sequence of instructions that makes it *look at the signs* of the two original numbers in the display and operand registers. Further instructions lead the controller through some simple logic to *figure out what the sign of the result should be*. In the case of adding and subtracting, this also requires seeing which of the two numbers is larger. The controller looks at signs the same way it compares the size of numbers, by the *comparison* function.

For example, as shown in *Figure 8-16*, to check the sign of the number in the display register, the controller waits until the sign digits (basket eight) are at the register outputs. Then it *subtracts the display-register sign bit from zero*. (The zero is provided by the "constant generator" subsystem.) A resulting borrow bit of 1 tells the controller that the display-register sign is negative. (In case you're curious, the subtraction process to generate the borrow bit is summarized in *Figure 8-16.*)

The controller then makes a note of this sign in the flag register—along with notes on the sign of the number in the operand register and on which of the two numbers is larger. After following instructions that lead it to figure out what the sign of the result in the accumulator register should be, it routes a 1 or 0 from the constant generator to the sign-digit position in the accumulator register. As you can see, the instructions make the controller follow much the same steps you would in working out a problem with pencil and paper.

**Figure 8-16.
A Subsystem for
Determining the Sign of
a Number**

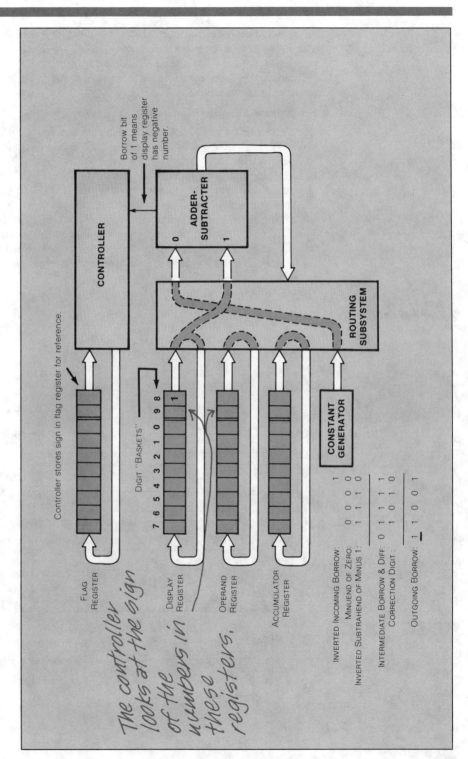

(Actually, a real calculator would most likely *not* store signs in the three number registers, but would simply keep track of the three signs *permanently in the flag register*. But for our initial understanding of calculators, it's more comfortable to think of the signs being stored right along with the numbers!)

How does the flag register work?

Since we've brought up the flag register again, let's see how it might work. *Figure 8-17* shows a design that would be compact and convenient. Here, we've arranged for the flag register to consist of two recirculating ten-bit shift registers labelled "A" and "B," like the four that make up each of the three number registers. "The flag register" and all three number registers are made as a single *mass-memory unit* containing *fourteen 10-bit shift registers*, all shifting in step together. Thus, one little "basket" at a time is accessible by the controller as the "digit basket" are assessed by the routing subsystem. Each basket contains two "flag bits" that can be read or written as the basket passes through the controller.

**Figure 8-17.
A Flag Register**

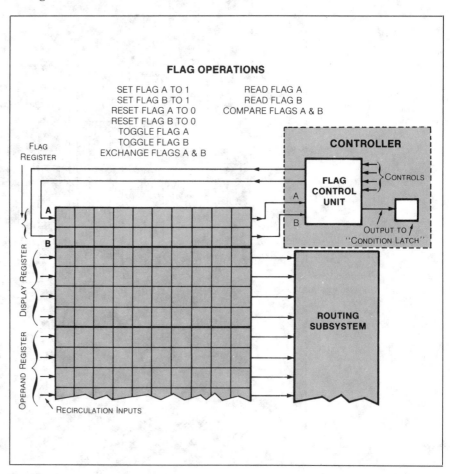

The flag register contains bits whose state indicates that particular operations have occurred or particular conditions have been met or not been met as the system operations proceed.

The reading and writing (and recirculation) of flags are handled by a "flag control unit" in the controller subsystem, shown in *Figure 8-17*. Shown above is a possible list of different "flag operations" that can be performed on the two flag bits in one basket by appropriate signals to the control unit.

Each of these operations would be done in response to an instruction word from the microprogram memory. The bit resulting from a "flag test" operation (reading or comparing flags) is stored in a one-bit storage unit in the controller, called the "condition latch." A later instruction would tell the controller to do one thing if the condition latch has a "1," and another if it has a "0." You'll need to keep this in mind as we study the subject of *programming* a system, in the next chapter.

How does the routing subsystem work?

We've mentioned the routing subsystem so often that we may as well see how it works too. It mainly consists of *data selectors*, which might be arranged as shown in *Figure 8-18*. Here, for simplicity, we're only looking at the part of the subsystem that's associated with the display register and one input to the adder-subtracter.

Signals are routed through the system by using data selectors to select the paths.

Each data selector has a number of four-bit inputs and one four-bit output, each for a decimal digit. One four-bit input digit to each data selector is selected for transmission to the output by control signals from the controller. As an example, broken lines in *Figure 8-18* show paths selected by the controller (out of many possible paths) to recirculate digits in the display register and also route them to one input of the adder-subtracter.

A feature is included in *Figure 8-18* that we haven't discussed before – namely, a method for *shifting* stored digits to the right or left among the ten baskets in each number register. Remember, we said back in Chapter 1 that before entering a new digit from the keyboard into the far right end of the display register, we have to *shift* the stored digits one step to the left. And to "line up the decimal points" before adding or subtracting, we may have to shift numbers to the left or right. So let's see how this shifting is done.

How are digits shifted right or left in the registers?

This shifting is done by controlling the "shift-control" data selector at the left in *Figure 8-18*, while the "load control" data selector is set to *recirculate* data. During one full recirculation of the register, the "shift-control" selector causes either *one shift to the left* (by selecting the upper input), *no shift at all* (by selecting the middle input), or *one shift to the right* (by selecting the lower input).

**Figure 8-18.
A Calculator Routing
Subsystem**

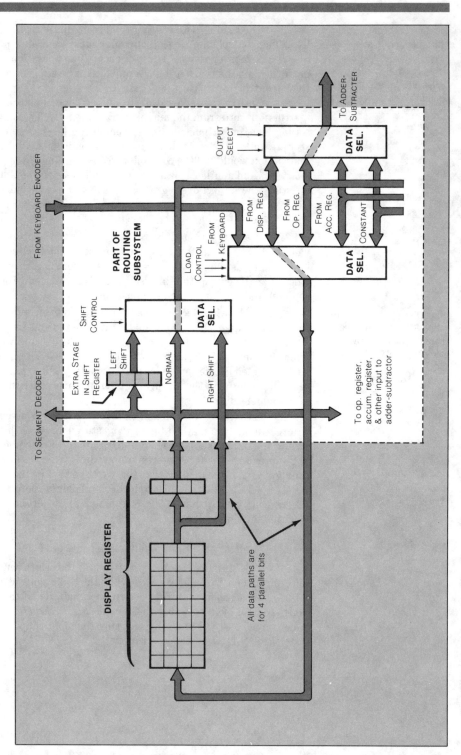

Shifting left or shifting right is controlled by selecting a data path that either adds a stage to the path (causing a delay) or eliminates a stage (causing an advance).

The "left-shift" path simply *adds one extra stage* to the shift register, thus *delaying* the recirculating digits one step behind their former positions. And the "right-shift" path *"short-circuits"* the recirculation path by one stage, by taking data from the *next-to-the-last stage* of the display register. This advances the digits one position *ahead* of the positions they had before.

There are some other details that must be attended to (such as switching back to the normal "no-shift" path before the sign and decimal-point digits get shifted, and resetting the two "selectable" stages to zero before cutting them into the path). But this should give you a general idea of how data in a recirculating shift register can be shifted right or left among the circulating digit baskets.

Where do we go from here?

In our process of studying how digital systems are put together to perform various functions, we have pretty well explained how nearly all of the parts of our calculator work. However, we know that the calculator is a "variable-programmed" system, in which sequences of instructions taken from a memory exert a "master control" over various hard-wired subsystems. We really haven't said very much about how stored programs work yet, so we haven't discussed the calculator's *operations that depend on programming*—such as how multiplication and division are performed, and so forth.

These topics will be taken up in the next chapter, as we discuss *programmed* digital systems. Our understanding of how the calculator works will lead us directly to a comprehension of microprocessors and computers, which are in many ways the most sophisticated of all digital systems.

Quiz for Chapter 8

1. The more bits you process at one time, the faster the job gets done, but:
   a. The lower the accuracy.
   b. The more circuitry required.
   c. The more total bits you have to process.
   d. The more separate steps required.

2. The operation of a "hard-wired" system or sub-system is determined by:
   a. The way the hardware is put together.
   b. The way the hardware is controlled by instructions stored inside the unit.
   c. The software in the unit.
   d. All of the above.

3. Typically, the subsystems or parts of a variable-programmed system are themselves:
   a. Variable-programmed.
   b. Hard-wired, subject to master control.
   c. Completely independent of master control.
   d. All of the above.

4. What is illustrated by the example of having the decoder subsystem's two latches dependent on different clock phases for their operation?
   a. "Control" signals can't always be depended on to say *exactly* when an action is to happen.
   b. The exact timing of a controlled action can be very important.

   c. In a "synchronous" system, most actions must wait for the right clock signal.
   d. All of the above.

5. In a typical "variable-programmed" system, what part provides the step-by-step sequencing and therefore corresponds to counters in hard-wired systems?
   a. The stored instructions.
   b. The decoder networks of the controller.
   c. The instruction register.
   d. The program counter.

6. What is the main advantage of variable-program control over hard-wired control?
   a. It's typically faster.
   b. It's easy to change the program by storing a different set of instructions.
   c. The system is typically smaller.
   d. All of the above.

7. If a sizeable piece of hardware is needed for a certain function, you'll get more for your money if:
   a. The hardware is dedicated to one job.
   b. The hardware is shared among several jobs as a "multipurpose" unit.
   c. Bits are processed in smaller groups.
   d. Variable-programmed master control is used.

(Answers in back of the book)

Programmed Digital Systems

In the last chapter, we were introduced to the idea of "programmed" digital systems. We learned that such systems are made of a number of "hard-wired" subsystems, which work together under the control of a "variable program." Thus, the "hardware" can be made to do any of a number of different jobs, according to the design of the "software," meaning the stored instructions that are followed one at a time, step by step.

In this chapter, we will learn more about programmed systems—how they work, how various kinds differ from one another, and how they can be used in a wide variety of applications. The *reason* we're spending an entire chapter on this subject is that more and more digital systems are using variable-programming rather than being completely hard-wired. It seems that programmed systems are the wave of the future in digital electronics.

Why are programmed systems "the wave of the future"?

Digital semiconductor circuitry has evolved from individual devices that must be wired together to extremely complex integrated circuits (VLSI) that have a complete system in one IC package.

The reason why more and more digital systems are using the variable-programmed approach is summarized in *Figure 9-1*. This figure shows several stages in the evolution of digital semiconductor circuitry, beginning in the early 1950's with circuitry made entirely of "discrete" (separate) devices.

By the early nineteen seventies, it became possible to put the main digital circuitry of *complete systems in just one integrated circuit.* (Large systems might instead have each *subsystem* or major part in a single IC.) This level of integration is commonly called "very-large-scale integration" or "VLSI," as opposed to earlier small, medium, and large-scale integration. These levels of complexity are explained in *Figure 9-1*.

However, no two systems are exactly alike. So unless you're going to make and sell a system in very large quantities (say, several hundred thousand systems), it typically won't pay to cram all its main circuitry into a single highly specialized IC. This is because it would cost a great deal to design the IC and put it into production. Thus, integrated circuits aren't economical unless they're mass-produced in large numbers.

The prohibitive unit cost of manufacturing small quantities of specialized VLSI circuits, led to the development of standard, mass-produced low-cost ICs that can be programmed to perform a variety of tasks.

Figure 9-1 indicates this fact for the early 1970's. On the one hand, we see that there will be a few systems sold in large enough quantities to use specially-designed VLSI circuits. But on the other hand, an increasing number of systems will use *standard, mass-produced* ICs (or sets of just a few ICs) of VLSI types that can be *programmed to serve a wide variety of systems.* For example, the same IC could possibly be programmed as the main circuitry of either the digital autopilot that we saw in *Figure 5-11*, or of a calculator. (Different accessory units would be used in each application, of course.)

**Figure 9-1.
Evolution of Digital
Semiconductor Circuitry**

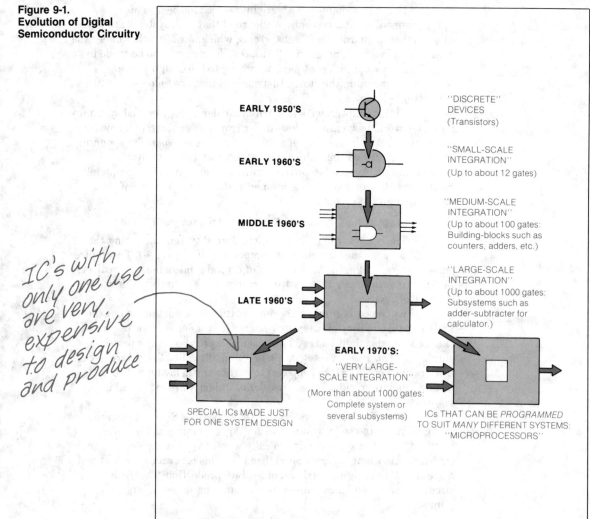

IC's with only one use are very expensive to design and produce

A number of different kinds of "general-purpose programmable" ICs and sets of ICs are available from semiconductor manufacturers. Many of these fall in a group called "microprocessors," which we will be studying later in this chapter. Because they can be used in so many different systems, microprocessors can be manufactured in *large enough quantities* for their costs to be as low as possible.

The story is much the same for systems that are too large to be made of a single microprocessor, or even several microprocessors. Rather than designing such a large digital system from scratch, it is usually more economical to buy a mass-produced *general-purpose programmable* system that's large enough for the job. Such larger systems are typically called a "computer." However, as we will see, the trend in digital electronics is to break large jobs up into parts that can be handled instead by several smaller programmed units instead of using a single computer. The smaller units (which may contain microprocessors) are typically small computers called minicomputers or even smaller ones called microcomputers.

How can one programmed system work in so many applications?

The reason that a programmable system can be used in a wide variety of tasks is that complex, specialized tasks are broken down into small, simple, unspecialized component parts.

Of course, the big question that immediately comes to mind is, "How in the world can just *one* set of hardware—whether it's a microprocessor, microcomputer, minicomputer, or computer—possibly provide the main circuitry for a *wide variety* of different systems?" That's the main question which this chapter is intended to answer.

In simple terms, the answer is that the systems use the method we have seen again and again in digital electronics. They *break down the big, complicated, specialized jobs into small, simple, unspecialized jobs.* By performing a large number of various small jobs—perhaps over and over again on various pieces of information—they get the big jobs done.

To begin getting an idea of just *how* this breaking-up of large jobs is done, let's turn once more to our example calculator, because it is a system that can be programmed to perform a rather wide variety of jobs. As we will see, it's quite a bit more specialized than typical microprocessors and computers, but it does have many features in common with such systems.

To be specific, let's see how several "big, complicated, specialized" jobs are broken down by the calculator and performed as a series of "small, simple, unspecialized" jobs. These small jobs are ones that can be handled by the calculator's main processing unit (the adder-subtracter), together with the routing and shifting processes that the routing circuitry can handle. The calculator handles its "big" jobs in *essentially the same way* as most microprocessors and computers, as we will see later.

The first example we'll look at is *multiplication* of numbers. We'll spend quite a bit of time on this example, in learning the basic concepts of programming. Then we'll quickly consider a couple of additional examples.

What's an "algorithm," and what does it have to do with multiplying?

The first thing to understand about solving any numerical problem—including multiplication—is the "algorithm" used for the process. An algorithm is a *logical plan* for solving a problem by a series of steps. It's the *idea* of a way to break a "big" problem into "smaller" problems. Although we haven't used the word "algorithm" before, we have already seen algorithms at work in the way hard-wired logic networks add and subtract numbers. Furthermore, the methods we learned in grammar school for adding, subtracting, multiplying, and dividing are "algorithmic" methods.

The algorithm our example calculator uses for multiplication is based on the ordinary algorithm for multiplying decimal numbers with pencil and paper. *Figure 9-2a* shows how this algorithm works in the example of multiplying 23.1 times 9.64. (You may remember from elementary school that 23.1 is the "multiplier" and 9.64 is the "multiplicand.") As you know, this customary algorithm involves calculating three "partial products," one for each digit of the multiplier, other than zeros. These partial products are then added to get the final product.

> An algorithm is a logical and clearly defined set of instructions for problem-solving. It is used to break down large problems or calculations into simple, smaller, more manageable parts.

**Figure 9-2.
An "Add-and-Shift"
Algorithm for
Multiplication**

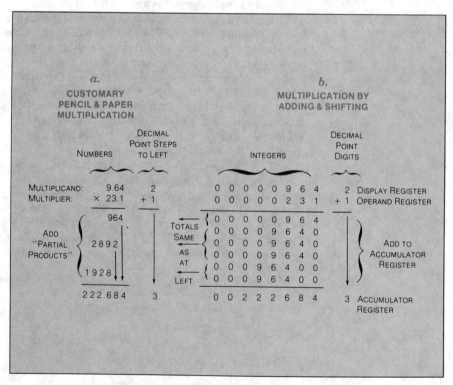

To calculate the decimal-point position in the final product, we count the steps to the decimal point from the right end of the multiplicand and multiplier (2 and 1 in *Figure 9-2a*). We *add* these steps together, getting a total of 3 steps from the right end for the decimal point in the product. Thus, the answer is 222.684.

What algorithm does the calculator use for multiplication?

Figure 9-2b shows the same problem as it would be solved by the algorithm used in the calculator. It works the same way, except that the effect of each partial product is created by *adding* the multiplicand the required number of times. For each of these multiple additions, the multiplicand is *shifted* to the left the proper number of steps.

The algorithm for multi-
plication in the calculator
involves only addition and
shifting of digits.

Notice in *Figure 9-2b* that the multiplier, multiplicand, and product are written in the way they are stored in the number registers in the calculator. The main part of each number is called the "integer," which means a number without a decimal point. For each integer, there is a decimal-point digit, showing how many steps to the left the decimal point goes. This is as we discussed it with respect to *Figure 8-11*.

To handle this problem in the calculator, we start with the multiplicand in the display register, the multiplier in the operand register, and an integer of zero (00000000) in the accumulator register. To this zero, we add the multiplicand integer (0000964) *once*, because the "ones" place in the multiplier integer is 1. The sum of 00000964 goes back into the accumulator.

Then we *shift* the multiplicand integer one place to the left in the display register, making it 00009640. We add it to the accumulator integer *three times*, because the "tens" place in the multiplier is 3. Each time we add, the sum goes back into the accumulator.

We shift the multiplicand left once more, and add the resulting 00096400 into the accumulator *two times*, according to the 2 in the "hundreds" place of the multiplier. That leaves our product integer of 00222684 in the accumulator register.

Finally, we add the decimal-point digits (DPDs) in the display and operand registers (2 plus 1), and put the sum of 3 in the DPD position in the accumulator. This completes the multiplication. We now have our answer in the accumulator register, in the form of an integer of 00222684 and a DPD of 3. The full multiplication took six additions and two shifts.

So that's how the calculator's multiplication algorithm works. To dig deeper into the subject, let's look at the actual instructions that our example calculator might follow as it performs a programmed multiplication. That is, let's look into the program itself. This will show us how *software* puts *hardware* through its paces. As we go through this exercise, you will be recalling and building on just about everything you have learned in various earlier chapters about how our example calculator works.

How would the multiplication algorithm be programmed?

A flowchart is a diagram or map of a program. It shows what steps must be performed and the order in which they are to be performed.

The best way to understand the working of a program for any system is to look at a "flow chart" of the program. *Figure 9-3* is a type of flow chart that outlines the general procedures for carrying out the calculator multiplication algorithm we have been discussing. As you can see, a flow chart is a sort of "road map" that shows various procedures as blocks, with arrows indicating the paths the system may take in going from one procedure to another. This chart is just a diagram of the algorithm as we discussed it, except that the DPD of the product is calculated at the beginning rather than the end, and we are including a block for some "final details" that need to be attended to.

Figure 9-3 will be useful for reference in studying a more *detailed* flow chart shown in *Figure 9-4*. Here, each block represents an actual *instruction* stored in the microprogram memory. For the sake of illustration, we're supposing that these instructions are stored at addresses number 62 through 77, as you see written above the blocks in *Figure 9-4*. Written in each block is a short description of what the instruction makes the calculator do.

**Figure 9-3.
A Multiplication
Algorithm Flow Chart**

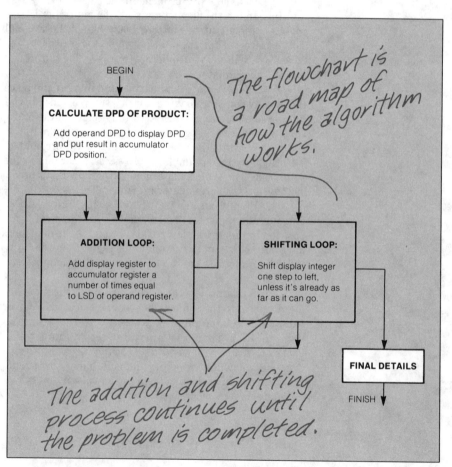

BEGIN

CALCULATE DPD OF PRODUCT:

Add operand DPD to display DPD and put result in accumulator DPD position.

The flowchart is a road map of how the algorithm works.

ADDITION LOOP:

Add display register to accumulator register a number of times equal to LSD of operand register.

SHIFTING LOOP:

Shift display integer one step to left, unless it's already as far as it can go.

The addition and shifting process continues until the problem is completed.

FINAL DETAILS

FINISH

**Figure 9-4.
A Multiplication
Microprogrammed
Routine**

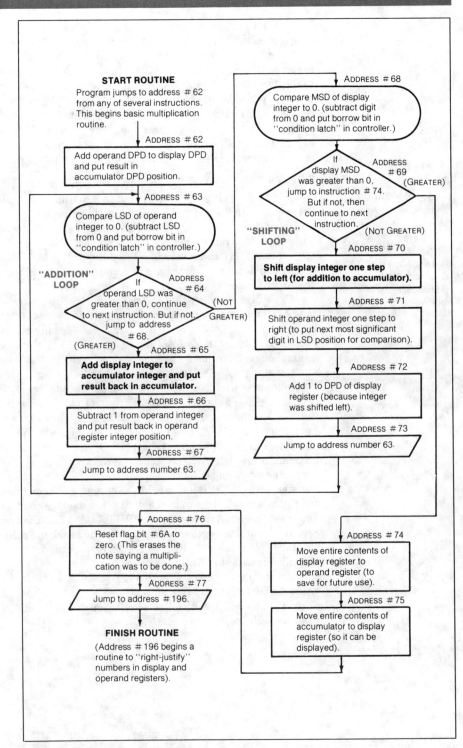

START ROUTINE

Program jumps to address #62 from any of several instructions. This begins basic multiplication routine.

ADDRESS #62

Add operand DPD to display DPD and put result in accumulator DPD position.

ADDRESS #63

Compare LSD of operand integer to 0. (subtract LSD from 0 and put borrow bit in "condition latch" in controller.)

"ADDITION" LOOP

ADDRESS #64

If operand LSD was greater than 0, continue to next instruction. But if not, jump to address #68.

(NOT GREATER)

(GREATER)

ADDRESS #65

Add display integer to accumulator integer and put result back in accumulator.

ADDRESS #66

Subtract 1 from operand integer and put result back in operand register integer position.

ADDRESS #67

Jump to address number 63.

ADDRESS #68

Compare MSD of display integer to 0. (subtract digit from 0 and put borrow bit in "condition latch" in controller.)

ADDRESS #69

If display MSD was greater than 0, jump to instruction #74. But if not, then continue to next instruction.

(GREATER)

(NOT GREATER)

"SHIFTING" LOOP

ADDRESS #70

Shift display integer one step to left (for addition to accumulator).

ADDRESS #71

Shift operand integer one step to right (to put next most significant digit in LSD position for comparison).

ADDRESS #72

Add 1 to DPD of display register (because integer was shifted left).

ADDRESS #73

Jump to address number 63.

ADDRESS #76

Reset flag bit #6A to zero. (This erases the note saying a multiplication was to be done.)

ADDRESS #77

Jump to address #196.

FINISH ROUTINE

(Address #196 begins a routine to "right-justify" numbers in display and operand registers).

ADDRESS #74

Move entire contents of display register to operand register (to save for future use).

ADDRESS #75

Move entire contents of accumulator to display register (so it can be displayed).

This group of instructions represents a particular type of program called a "microprogrammed routine" (or "subroutine"). The prefix "micro" means that this is a type of program that's stored *permanently* in some sort of ROM (read only memory). (We'll come back to this fact later.) And a "routine" (or subroutine) is a sort of "small" program which is part of a larger one. For example, you may remember that in Chapter 1, we spoke of the "idle" routine," the "add" routine, and so forth. What *Figure 9-4* represents is the calculator's *multiplication* routine (in simplified form). Whenever the controller must do a multiplication, it is sent to instruction number 62 to begin the routine.

What facts must be recalled to understand this routine?

In following the operation of this routine, remember that as we first noted with respect to *Figure 1-7*, each instruction that the controller draws from the microprogram memory governs what happens during one "instruction cycle." This is the period of time (about 100 microseconds—100 millionths of a second) during which one scan line is active, and during which the number and flag registers complete exactly one recirculation.

During an instructon cycle, bits are moved to other registers or shifted within a register, or flag bits are set or reset to accomplish the various operations of a routine, such as addition, subtraction, or comparison.

Also remember that during one instruction cycle, all the digits in two number registers can be passed to the adder-subtracter for addition, subtraction, or comparison (*Figure 7-5* will remind you of how these parts fit together.) Also, all the digits in a register may be moved to another register during one instruction cycle. Or the "integer" digits in a register may be shifted one step to the left or right. Or instead, the controller may perform one of the "flag operations" we noted in *Figure 8-17*, either changing or "testing" one or both of a pair of flag bits.

Furthermore, keep in mind that the adder-subtracter's operations can be performed only on *certain digits* in the registers. For example, a new decimal-point digit (DPD) can be inserted into the accumulator register without touching the sign or integer digits. The controller would cause this to happen by recirculating the accumulator register until the DPD appears at the output, and then quickly switching the right data selectors to load a digit from, say, the output of the adder-subtracter. After this new DPD is loaded into the accumulator register input, the register would be switched back to "recirculate." All this takes place in one instruction cycle, and is ordered by certain bits in the coded instruction being followed.

Similarly, the integer in a register may be shifted right or left without shifting the DPD or sign digit. The controller does this by switching the "shift-control" selector *(Figure 8-18)* to one of the "shift" positions *only while the integer is passing through.*

What must be understood about selecting the next instruction?

Instructions normally will advance by adding one to the current program counter address.

It's particularly important to understand how the controller *decides which instruction to select* for the next instruction cycle. Most of the instructions shown in *Figure 9-4* cause the program counter *(Figure 8-5)* to *add* 1 to the current address. This causes the *next instruction in line* in the microprogram memory to be selected for the next instruction cycle. This process of simply going to the next instruction is represented in *Figure 9-4* by vertical arrows from one instruction down to the next.

Placing an instruction address in the program counter, without having conditions to satisfy, is called an unconditional jump instruction.

Then there are "unconditional jump" instructions, which are represented in *Figure 9-4* by trapezoidal (slanted) boxes. These instructions simply provide a *new address*, different from that obtained by adding 1, for the controller to put into the address register. Nothing else happens during one of these instruction cycles. The number and flag registers simply recirculate without being changed, while the specified new instruction is fetched for the next instruction cycle. As you can see in the flow chart, an unconditional jump instruction makes the controller "jump" to another point in the routine.

Finally, there are the *"conditional jump"* instructions, which are shown as diamond-shaped boxes in *Figure 9-4*. These instructions also contain a new address for a "jump." But they make the controller look at the *condition latch* (mentioned in *Figure 8-17*) and *decide whether or not to jump*, depending on whether a 1 or a 0 is found there.

For a conditional jump instruction, certain conditions must be satisfied before the instruction address is placed in the program counter.

For example, consider the conditional-jump instruction at address number 64 (in the addition loop in *Figure 9-4*). What this instruction really tells the controller is, "Jump to address number 68 if you find 0 in the condition latch." For another example, instruction number 69 says, "Jump to address number 74 if you find 1 in the condition latch." If the controller doesn't find the bit it's looking for in the condition latch, it simply causes 1 to be added to the current address as usual. Nothing else takes place during an instruction cycle governed by a conditional-jump instruction.

Now, where does the bit in the condition latch come from? It's loaded in during the *preceding* instruction cycle, from a source that's specified in that instruction. In the examples we just mentioned, the preceding instructions (oval blocks in *Figure 9-4*) were for comparison operations by the adder-subtracter, and the "borrow" bit from the carry-borrow storage unit *(Figure 8-15)* was loaded into the condition latch. If these instructions had been for flag test operations instead, then the result of the flag test would have been put into the condition latch, as we noted in *Figure 8-17*.

Conditional-jump instructions are this particular system's way of making a *branch point* in a routine. This provides a method for *deciding what to do next*, based on the results of previous operations. Branching provides the main *decision-making capability* in a program.

How can we check the operation of the multiplication routine?

Now if you can keep all these capabilities of the calculator in your head, you can proceed to follow the operation of the multiplication routine in *Figure 9-4*. To do this, you may use the example multiplication problem from *Figure 9-2* (23.1 times 9.64).

As you go, you may refer to *Figure 9-5*, which is a record of what happens in the calculator as these two numbers are multiplied. Each horizontal line in *Figure 9-5* shows first (left) the addresses of one or more instructions in the flow-chart that are followed by the controller, one after another. Then the rest of each line shows the decimal-point digit and integer in each of the three number registers, as they would appear *after* the instructions to the left are performed. The numbers that *change* as a result of these instructions are shown in bold print in *Figure 9-5*.

As we mentioned earlier in discussing the multiplication algorithm, this routine begins with the multiplicand and multiplier *already loaded into the appropriate registers*. (You'll see this on the top line in *Figure 9-5*.) These numbers were put there by whatever routine was performed before this one. There may be any number of other routines in the microprogram memory that contain conditional or unconditional jumps to the beginning of this multiplication routine.

What are some important details of this routine?

You'll note that the multiplication routine contains an instruction (address 74) to *preserve the multiplicand* (9.64) after the multiplication is complete. This is done by moving the multiplicand from the display register into the operand register, *before* the product is moved to the display register by the next instruction. This is in case the multiplicand needs to be used for further calculations in a "chain" of calculations.

Instructions are required to save the multiplicand, to remove insignificant zeroes, to adjust the DPD as needed, and to set or reset flag bits.

Also note that the last instruction of the routine (address 77) is an unconditional jump to another routine. This other routine (not shown) "right-justifies" the multiplicand and product integers. That is, it checks for any *zeroes* at the right of these integers, and *shifts* the integers to the right until there are no zeroes there. Each time an integer is shifted one place to the right, of course, its DPD must be reduced by 1, to keep track of the decimal point accurately.

Another detail worth noting is that instruction 76 resets a certain bit in the flag register in *Figure 8-17* (which we're supposing is bit number 6A—the "A" bit in the sixth position of the ten possible positions or "baskets" for a pair of flag bits). To "reset" means change to zero. This step erases a note saying that a multiplication was to be performed. This is a sample of how the controller uses the flag register to help keep track of what's being done and what needs to be done.

Situations such as a register overflow which could arise during multiplication also must be handled by instructions in the routine.

Finally, you should be aware that even though the routine in *Figure 9-4* may look complicated, it's actually *highly simplified* for purposes of instruction. For one thing, a complete multiplication routine would also include instructions to check for whether the product has become too big to be handled. If that happens, it's called an "overflow" problem. Further processing must be stopped and an "overflow" symbol must be illuminated in the calculator display. There are several other possible problems that could arise that must be provided for by appropriate additional branches. But the simplified routine we have seen should be sufficient to introduce you to the main ideas of *programming* a system.

Figure 9-5.
Sequence of Instructions for Multiplication Algorithm

ADDRESS NUMBERS OF INSTRUCTIONS FOLLOWED, RESULTING IN NUMBERS TO RIGHT	DISPLAY REGISTER	OPERAND REGISTER	ACCUMULATOR REGISTER
	D M L / P S S / D D D / INTEGER	D M L / P S S / D D D / INTEGER	D M L / P S S / D D D / INTEGER
BEGIN ROUTINE	2,00000964	1,00000231	0,00000000
62	2,00000964	1,00000231	3,00000000
63, 64, 65	2,00000964	1,00000231	3,00000964
66	2,00000964	1,00000230	3,00000964
67, 63, 64, 68, 69, 70	2,00009640	1,00000230	3,00000964
71	2,00009640	1,00000023	3,00000964
72	3,00009640	1,00000023	3,00000964
73, 63, 64, 65	3,00009640	1,00000023	3,00010604
66	3,00009640	1,00000022	3,00010604
67, 63, 64, 65	3,00009640	1,00000022	3,00020244
66	3,00009640	1,00000021	3,00020244
67, 63, 64, 65	3,00009640	1,00000021	3,00029884
66	3,00009640	1,00000020	3,00029884
67, 63, 64, 68, 69, 70	3,00096400	1,00000020	3,00029884
71	3,00096400	1,00000002	3,00029884
72	4,00096400	1,00000002	3,00029884
73, 63, 64, 65	4,00096400	1,00000002	3,00126284
66	4,00096400	1,00000001	3,00126284
67, 63, 64, 65	4,00096400	1,00000001	3,00222684
66	4,00096400	1,00000000	3,00222684
67, 63, 64, 68, 69, 70	4,00964000	1,00000000	3,00222684
71, 72	5,00964000	1,00000000	3,00222684
73, 63, 64, 68, 69, 70	5,09640000	1,00000000	3,00222684
71, 72	6,09640000	1,00000000	3,00222684
73, 63, 64, 68, 69, 70	6,96400000	1,00000000	3,00222684
71, 72	7,96400000	1,00000000	3,00222684
73, 63, 64, 68, 69, 74	7,96400000	7,96400000	3,00222684
75	3,00222684	7,96400000	3,00222684
76, 77	3,00222684	7,96400000	3,00222684

What aspects of programming have been illustrated?

The programmer must anticipate and provide instructions for every possible situation that may arise. The program must be designed so that the minimum number of instructions and minimum amount of time are required to accomplish the task.

The principles illustrated in our microprogrammed multiplication routine apply to the programming of *any* system that has the capabilities of conditional and unconditional jumps or branching. This includes microprocessors and computers.

First, you've seen the idea of a *routine* (or subroutine) that is called into action by some other part of the overall program. You've also seen the idea of a *"loop,"* consisting of a series of instructions that is followed over and over again until some condition is met that *branches the system out of the loop.* (Our routine has an "addition" loop and a "shifting" loop. The actual addition and shifting instructions at addresses 65 and 70 are shown in bold type in *Figure 9-4* for emphasis.)

Perhaps most important, you've had a look at the kind of *fine details* that must be considered in designing a program to do a certain job. Algorithms that look simple at first glance may turn out to require a surprising number of instructions with all manner of branches and loops.

Of course, the reason for the fine details is that an electronic system *can't think for itself.* The software designer must do all the thinking that's required—and *all in advance.* Every possible situation that could occur must be thought about and provided for by proper instructions. There are so many things that can go wrong in even the simplest programs that they typically don't run right the first few times they're tried. They've got to be carefully "debugged" (made correct).

Furthermore, as you can well imagine, the program must be *efficient* in terms of the *time and number of instruction cycles* that are involved in following it. Our example routine (which is not intended to be very efficient) takes *seventy-nine* instruction cycles just to perform the example multiplication. (You can count the instructions in the left column of *Figure 9-5.*) As we have noted before, *time* is the big limitation of a programmed system's capabilities. It takes *more time* to do a big job as a series of steps than all at once. The fewer the steps required, the faster the job gets done.

And finally, a program must be efficient in terms of the *number of instructions it contains.* After all, in the case of our example calculator's microprogram memory, there is a definite limit to the number of instructions it can hold.

These, then, are some important things that nearly all programs have in common. You should keep them in mind as we proceed to study other calculator routines, and later on as we look into microprocessors and computers.

How would the calculator handle division?

The algorithm for division in the calculator involves subtraction, comparison and shifting of digits. The comparison prevents a negative remainder.

Since you've seen how our example calculator handles multiplication, you may well be curious as to how it might perform division. So let's take this as our second example of how programmed systems perform complicated jobs as sequences of simpler jobs. It will be enough for us to look at the principle of the *algorithm* involved, without examining the actual microprogrammed routine.

The division algorithm works by "reversing" the add-and-shift process used for multiplication, and by "*subtracting*-and-shifting" instead. This procedure is very much like the "long division" algorithm you probably learned in grammar school. *Figure 9-6* shows this similarity in the example division of 1,058 by 46.

As you see in *Figure 9-6b*, we begin with the "dividend" (1,058) in the display register and the "divisor" (46) in the operand register. (We're ignoring decimal points in order to simplify your learning of the main idea.) The basic procedure is to *shift* the divisor to certain positions and *subtract* it from the display register as many times as we can, without getting a negative remainder. (Each subtraction is preceded by a *comparison*, to make sure the remainder won't be negative.) After each subtraction, the remainder goes *right back into the display register*, in the place of the original dividend.

Meantime, in the LSD (far right end) of the accumulator register, we *keep count of the subtractions*. When we've subtracted as many times as we can, we *shift* the divisor one step to the right and begin the process again.

**Figure 9-6.
A "Subtract-and-Shift"
Algorithm for Division**

The division algorithm is very similar to the one for multiplication except you subtract rather than add.

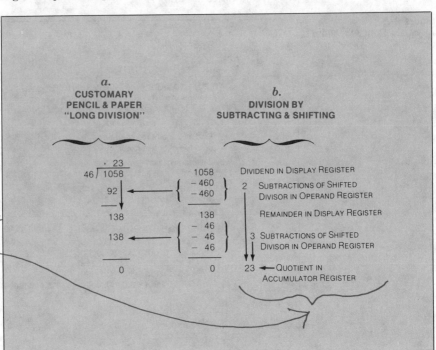

a.
**CUSTOMARY
PENCIL & PAPER
"LONG DIVISION"**

b.
**DIVISION BY
SUBTRACTING & SHIFTING**

```
        , 23
   46 ) 1058          1058        DIVIDEND IN DISPLAY REGISTER
        92          - 460      2  SUBTRACTIONS OF SHIFTED
                    - 460         DIVISOR IN OPERAND REGISTER
       138           138         REMAINDER IN DISPLAY REGISTER
       138          -  46      3  SUBTRACTIONS OF SHIFTED
                    -  46         DIVISOR IN OPERAND REGISTER
                    -  46
         0             0      23 ← QUOTIENT IN
                                  ACCUMULATOR REGISTER
```

Each time we shift and start subtracting from the remainder again, we *shift the accumulator register one place to the left*, giving us a new LSD of 0. That way, we begin counting subtractions again with another least-significant digit. The subtracting-and shifting process is stopped when a remainder of zero is produced, or when the accumulator register has been filled up all the way to the left end with "counter" digits. These digits, then, are the "quotient" we want. The decimal-point digit for the quotient would be figured out by a process that involves *subtracting* DPDs rather than adding as we did for multiplication.

Even in simplified form, a flow chart for a division routine would be more complicated than we would want to tackle, so we will not go into further details. What you've seen should be enough to give you the general idea of what's involved.

How are approximations used for more complicated calculations?

Higher mathematical operations are established by programmed sequences of the fundamental four operations of addition, subtraction, multiplication and division.

You're probably aware that many calculators can, at the press of a key, make calculations other than the "basic four" of addition, subtraction, multiplication, and division. For example, there are the mathematical "functions" of the square root of a number, the sine of an angle, and many others. Calculators and other programmed systems often perform such "higher mathematical" operations simply by the appropriate *programmed sequences of the basic four operations* that we have already learned about. These sequences involve calculating closer and closer *approximations* of the desired number, until the result is as accurate as it can be expressed by the limited number of digits used in the system for each number.

**Figure 9-7.
A Square Root Algorithm**

TO FIND SQUARE ROOT OF A NUMBER:
Call the number "N."
Use 2 as an approximation of the square root. Then:

1. Divide N by approximate root.

2. Find average of result and approximate root (add together and divide by 2).

3. If average is same as approximate root, you're as close as you can get to the square root.

4. If not, use average as new approximate root, and go back to *Step 1*.

EXAMPLE: N=9
Begin with approximation of: 2.000000000

$[(9 \div 2.000000000) + 2.000000000] \div 2 = 3.250000000$ (NOT SAME)]
$[(9 \div 3.250000000) + 3.250000000] \div 2 = 3.009615385$ (NOT SAME)]
$[(9 \div 3.009615385) + 3.009615385] \div 2 = 3.000015360$ (NOT SAME)]
$[(9 \div 3.000015360) + 3.000015360] \div 2 = 3.000000000$ (NOT SAME)]
$[(9 \div 3.000000000) + 3.000000000] \div 2 = $ **3.000000000 (SAME)**

For example, *Figure 9-7* explains an algorithm that finds the *square root* of a number by a series of closer and closer approximations. (The square root is what you would multiply by itself to get the number.) You begin with *any number* as an approximation of the square root—say, the number 2. Then you find a number that's a *closer* approximation of the square root by following the four numbered steps listed in *Figure 9-7*. By repeating this series of four steps, you can get as close to the true square root as the length of your calculator's numbers will allow. The example of *Figure 9-7* requires doing the steps 5 times on a calculator that handles 10 digits.

Complex functions are calculated by using substitute mathematical expressions that can be solved by computer. The Maclaurin Series for the sine function is an example.

For another example of a series-of-approximations solution of a function, *Figure 9-8* shows one of several possible ways to calculate the *sine of an angle*. This figure will remind you of what the sine of an angle is (namely, the length of the side of a right triangle opposite the angle, when the hypotenuse is 1).

Rather than describing an algorithm for this method, *Figure 9-8* shows a *series of algebraic terms* that form the approximate answer. (A "term" is something added or subtracted.) This series is called the "Maclaurin series" approximation for the sine of an angle. Each term (beginning at the left with the angle, Z, in radians) is just the term to its left multiplied twice more by Z and divided by the next two higher integers (1,2,3,4, etc.), with the sign changed. Four terms are shown, but the series is theoretically "infinite," meaning you can add and subtract as many terms as you need for accuracy in approximating the sine. (Eventually, the terms get smaller and smaller.)

Figure 9-8.
A Sine Function
Approximation

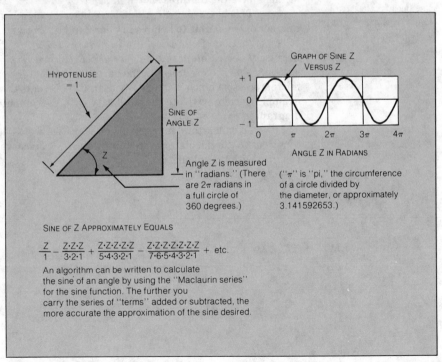

HYPOTENUSE
= 1

SINE OF
ANGLE Z

Angle Z is measured in "radians." (There are 2π radians in a full circle of 360 degrees.)

GRAPH OF SINE Z
VERSUS Z

+ 1

0

− 1

0   π 2π 3π 4π

ANGLE Z IN RADIANS

("π" is "pi," the circumference of a circle divided by the diameter, or approximately 3.141592653.)

SINE OF Z APPROXIMATELY EQUALS

$$\frac{Z}{1} - \frac{Z\cdot Z\cdot Z}{3\cdot 2\cdot 1} + \frac{Z\cdot Z\cdot Z\cdot Z\cdot Z}{5\cdot 4\cdot 3\cdot 2\cdot 1} - \frac{Z\cdot Z\cdot Z\cdot Z\cdot Z\cdot Z\cdot Z}{7\cdot 6\cdot 5\cdot 4\cdot 3\cdot 2\cdot 1} + \text{etc.}$$

An algorithm can be written to calculate the sine of an angle by using the "Maclaurin series" for the sine function. The further you carry the series of "terms" added or subtracted, the more accurate the approximation of the sine desired.

Where do we stand in learning about programmed systems?

By studying methods used by a calculator for mathematical operations beyond addition and subtraction, we have gained a general idea of how flexible and versatile a programmed system can be. We have seen some pretty impressive examples of the fact that just about any mathematical operation can be performed as a programmed series of simpler operations. We have seen how a stored *program* can put a system made of our familiar building-blocks through an incredible variety of different performances, much as a musical score makes the instruments of an orchestra play anything from Beatles to Beethoven.

However, as we noted earlier in this chapter, our calculator is not as versatile as the "general-purpose programmed systems" that we called the wave of the future—namely, computers (including mini and microcomputers) and their integrated-circuit cousins, microprocessors. This is because the calculator's circuitry was *specifically designed* for a rather basic calculator. So to complete the picture of programmed systems, let's look into a *truly general-purpose* programmed system. This will lead us into an understanding of microprocessors and computers.

What are the main features of a "general-purpose" programmed system?

Most of what we have seen in studying the calculator applies to general purpose, programmed systems because the calculator is a limited, specialized type of a general purpose system.

Figure 9-9 shows a very general block diagram that represents the main circuitry of *any system* of the type that we're calling "general-purpose." The details of such systems vary endlessly, but most of them have the features shown in *Figure 9-9*.

This block diagram could have been arranged in any of the number of ways. But to help you understand it, the arrangement is made to resemble that of our calculator in *Figures 1-6* or *7-1*. This is possible because *the calculator is just a specialized, limited version of this kind of system. Figure 9-9* represents the parts of a computer (or even a microprocessor) much more closely than it represents those of our example calculator. (Note the similarity with the computer block diagram in *Figure 7-15*.)

To understand how this general system works, let's see how it compares to the calculator system in *Figure 1-6*, since we're already familiar with the calculator.

**Figure 9-9.
A General-Purpose
Programmed Digital
System**

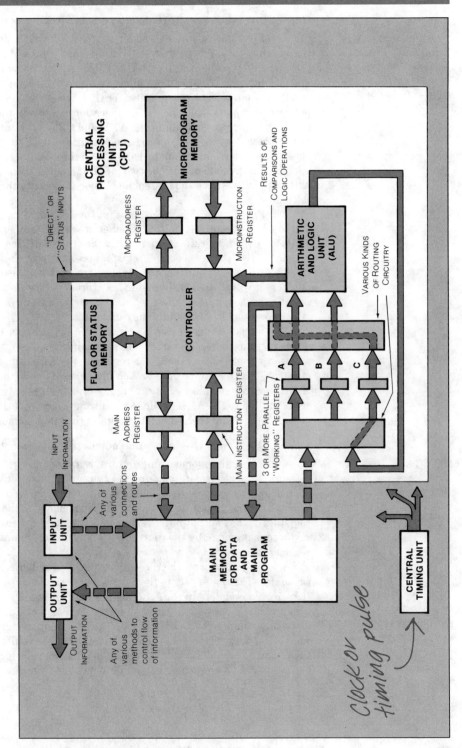

How do the "ALU" and its registers compare to the calculator?

The ALU not only performs the basic math functions and comparisons, but also performs logic gating operations and more extensive comparisons.

First, the calculator's adder-subtracter is just a simple form of the "arithmetic and logic unit" or ALU in *Figure 9-9*. ("ALU" is pronounced by spelling it, like A-L-U.) Most ALUs can do more than just add and subtract two incoming binary numbers and compare them for whether or not a certain one is greater than the other. They can also perform any of several *logic gating* operations on bits in the two binary numbers coming in—including the AND and OR operations. They also perform a wider range of *comparison* operations, such as "less-than-or-not" and "equals-or-not," and these comparisons can be applied to individual *bits* in the two input numbers. These additional ALU capabilities give the system much more flexibility than the calculator.

Furthermore, the *registers* that the ALU gets its inputs from and to which it routes its output number are *individual parallel registers*, called "working registers," rather than shift registers as in the calculator. (Only three working registers are shown in *Figure 9-9*, labeled A, B, and C, but there are typically more than this.) The binary numbers involved are considered to be "words," as we discussed in Chapter 7. The length of these words ranges from 4 bits in some microprocessors to 64 bits in some computers.

Various kinds of routing circuitry are typically associated with ALUs (perhaps in a configuration as shown in *Figure 9-9*) to move the words in various paths among the registers, the ALU, and the "main memory." As an example, the broken lines shown inside the two "routing circuitry" blocks indicate that words from registers A and B are being fed to the ALU, while the resulting word from the ALU goes into register C and then (in a later step) to the main memory.

How does the "main memory" unit compare to the calculator?

The main memory in a general purpose system is of the random access type, and access is by addresses for each word.

The main memory unit, in turn, is a random-access mass memory, with a separately-addressed storage location for each word. The controller causes words to be moved into the memory from the working registers or the "input unit," and out of the memory to the working registers or the "output unit." The memory address for each word involved is provided by the controller by way of the "main address register."

As you can see, the three recirculating number registers in our calculator *(Figure 7-5)* serve the purposes of *both the main memory and the working registers* in the general system *(Figure 9-9)*. In effect, the first and last stages of the shift registers serve as working registers, since they communicate directly with the adder-subtracter, through the routing circuitry. The remaining eight stages of each shift register, then, may be considered a sort of main mass memory, which is *serially accessed by shifting* rather than by random addressing.

How do the "input" and "output" units compare to the calculator?

Because it is general purpose, rather than fixed purpose, like a calculator, the input and output functions vary widely in a general purpose system. Typically, however, these functions must do some processing and control.

The "input unit" and the "output unit" in *Figure 9-9* represent functions that vary widely among specific programmed systems. These blocks are included in the figure to indicate that typically some sort of processing and control must be exerted on information moving into and out of the system. In our calculator *(Figure 1-6)*, the keyboard and keyboard encoder act as an input unit, and the segment decoder and display act as an output unit.

The pathway labelled "direct or status inputs" in *Figure 9-9* indicates that typically, certain external information is supplied directly to the controller, for it to use in deciding what to do. In the calculator *(Figure 1-6)*, this pathway corresponds to signals from the keyboard encoder that tell the controller when certain keys are pressed (as we discussed in Chapter 2). The calculator's "idle routine" causes it to look at each of the wires from the encoder, one at a time, and load the transmitted bit into the condition latch. After each of these "input-checking" steps, a conditional-jump instruction branches to an appropriate routine if the condition latch says the key was pressed.

In some systems—especially computers—the direct inputs to the controller are considerably more complicated than in the calculator. But the general principle remains the same.

How does the system's programming compare to the calculator's?

The instructions stored in main memory for a general purpose system typically direct the controller to perform many instructions contained in a permanent microprogram memory between the instructions from main memory.

Together with the random-access memory, the main feature that sets the general system in *Figure 9-9* apart from our calculator is this: The controller is not governed only by the routines stored permanently in the microprogram memory as in the calculator. In addition, the controller draws *instructions from a "main program"* stored in the main memory, by way of the "main instruction register" in *Figure 9-9*. The main program operates on the same principles we've seen for the microprogrammed routines, by the controller's selecting a series of stored instructions to follow. Typically, however, one "main instruction" makes the controller go through an *entire microprogrammed* routine, before the next main instruction is selected.

Probably the best way to think of this is in terms of a *programmable calculator*, which is a calculator in which the operator can *store a "programmed" series of keystrokes*. (Some programmable calculators even allow you to store unconditional and conditional jump instructions.) In running a program in one of these calculators, the stored instructions "press the keys automatically," in effect—according to the pattern you keyed into storage earlier. Each keystroke (whether finger-pressed or programmed) typically triggers a *microprogrammed* routine or series of routines. At the end of this microprogrammed process, then, the controller goes to the next main instruction it has selected, rather than to an "idle" routine as we discussed in Chapter 1 for our calculator.

Notice in *Figure 9-9* that the two registers associated with the microprogram memory are now labelled "*micro*address register," and "*micro*instruction register." The purpose or use of these registers is the same as we learned for our calculator. But the prefix "micro" is added to distinguish them from the *main* address and instruction registers. So we say that the *microprogram* memory stores *microinstructions* in locations identified by *microaddresses*.

The microprogram level of instructions controls the detailed, always-the-same operations which may control hard-wired functions. The main program which is different for different applications, calls the microprogrammed instructions through the main instructions.

Thus, as you can see, a general-purpose programmed system typically has *two "levels" of programming*, the main program and the microprogram levels. And as we have seen, the microprogrammed routines, in turn, exert control over subsystems and parts whose functions are *hard-wired* into their circuitry. So counting this lowest hard-wired level, we see *three levels of control in all*.

Each of these levels has its own general field of supervision, somewhat like fields of government at the local, state, and national levels, or perhaps more like the administrative, tactical, and strategic levels of command in an army. The hard-wired level is the circuitry itself, with its basic, unchangeable capabilities. The microprogrammed routines control fine, petty details of the system's operation—details such as the many instructions that must be followed to perform multiplication and division. Since such details are handled the same way every time (say, for every multiplication or division), the microprogram memory is *usually* a read-only type as in the calculator, although it may be a type that can be programmed by the user such as a PROM or EPROM, so that program changes can also be made at this level.

As you can see, the microprogrammed routines make life much easier for the person who develops the *main* software for a system. He can just throw in an instruction that says something like, "Multiply numbers from so-and-so addresses and put the answer in such-and-such address." At this, the microprogrammed routines take over and cycle the system through dozens or even hundreds of microinstruction cycles to do the job specified by just one main instruction.

We could go on and on in discussing this subject of programming. But let's stop with this introduction and move on to see how computers and their smaller counterparts, microprocessors and microcomputers, compare to the general picture of programmed systems in *Figure 9-9*.

How does microprocessor circuitry compare to this general picture?

The definition of what is included in a microprocessor varies quite a lot. The narrowest definition includes only the CPU while the widest definition includes everything in Figure 9-9.

There are any number of different IC chips and sets of chips on the market that can be called "microprocessors," and these systems differ considerably from one another. Furthermore, the name "microprocessor" is a very loose one, without a formal definition. Some systems called microprocessors would contain *all* the parts shown in *Figure 9-9*. And others would include only those in the large box in *Figure 9-9*, which is typically referred to as the "central processing unit" or CPU (pronounced by spelling, like C-P-U).

Some microprocessors do not use microprogrammed routines, while others use a microprogram memory on a chip apart from the controller. And some microprocessors are *completely microprogrammed* like our calculator, without any readily-changed main program. Finally, the routing and interconnections of words among the main memory, working registers, and ALU vary considerably among various micropressors.

Typical microprocessors use shorter words than larger computers - 4, 16, 32 bits compared to 16, 32 or 64 bits.

One of the main features that distinguishes microprocessors from systems called "computers" is that *the words are shorter*. Most microprocessors use from 4-bit to 16-bit words, while "computers" (including "minicomputers") have word lengths of 16 to 64 bits. We'll see other differences when we come to computers a little later in the chapter.

How can microprocessors be applied in a computer?

However, you should note that a computer may be *made from one or more microprocessors*. One way to do this is illustrated in *Figure 9-10*. Here, we see the general idea of using four microprocessor IC chips of a type called "bit-slice" to form the CPU for a "microcomputer" (a very small computer) that uses 16-bit words. (Only two of the four chips are shown, to save space in the picture.) Each bit-slice chip works on *four bits* of the 16-bit words being processed.

There is a class of computers called "bit-slice" computers that use multiple microprocessors to handle only a portion of a computer word, say 4 bits of 16. All microprocessors work in parallel to process the complete word.

During the addition or subtraction of 16-bit words, carry-borrow bits are passed from one 4-bit ALU to another. Similarly, when numbers must be shifted in the working registers for performing multiplication and division, bits pass between the registers of neighboring chips. Coordination of the efforts of the four chips is handled by control signals represented at the right of *Figure 9-10*. This arrangement shows you how versatile certain microprocessor chips can be. Such microprocessors are used for highest system performance, or to provide for instructions required for a very special system. These are called user-defined instructions because the user can make up his own instructions, different from those decided on by the microprocessor manufacturer.

**Figure 9-10.
A Bit-Slice 16-bit
Microcomputer CPU**

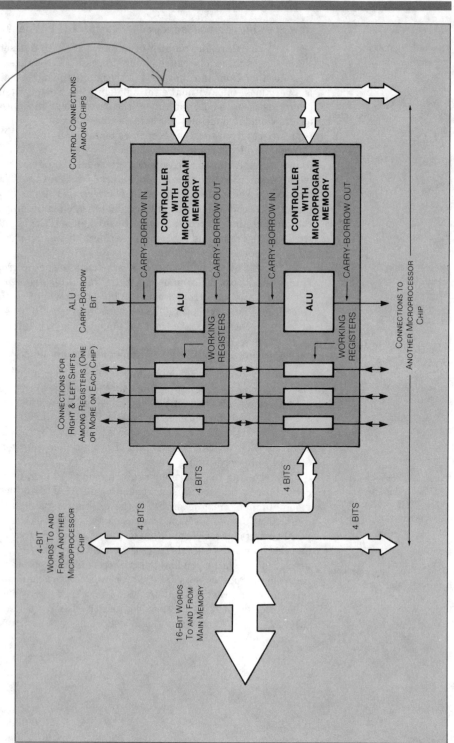

Timing and coordination comes from this line.

CONTROL CONNECTIONS AMONG CHIPS

CONTROLLER WITH MICROPROGRAM MEMORY

CONTROLLER WITH MICROPROGRAM MEMORY

CARRY-BORROW IN

CARRY-BORROW OUT

CARRY-BORROW IN

CARRY-BORROW OUT

ALU CARRY-BORROW BIT

ALU

ALU

WORKING REGISTERS

WORKING REGISTERS

CONNECTIONS TO ANOTHER MICROPROCESSOR CHIP

CONNECTIONS FOR RIGHT & LEFT SHIFTS AMONG REGISTERS (ONE OR MORE ON EACH CHIP)

4 BITS

4 BITS

4-BIT WORDS TO AND FROM ANOTHER MICROPROCESSOR CHIP

4 BITS

4 BITS

16-BIT WORDS TO AND FROM MAIN MEMORY

How can a microprocessor be used in a calculator?

On the other hand, a single 4-bit microprocessor chip that contains *all* the parts shown back in *Figure 9-9* could be microprogrammed to provide the main circuitry for a *calculator*. Since it would have provisions for a "main program" in addition to a microprogram, the calculator could be the *programmable* type.

The best way to view a microprocessor is as a programmable building block that can handle simple or complex tasks depending on how it is programmed.

The ALU would work on one pair of 4-bit BCD digits at a time. These digits would be drawn one at a time from the main memory and placed in working registers for processing by the ALU. The resulting "sum" or "difference" digit would then be shuttled back into the main memory. Eight cycles of this process would be required to add or subtract an 8-digit decimal integer. (The ALU would contain a one-bit latch like that in our calculator's adder-subtracter, for holding a carry-borrow bit.)

The microprogramming of such an 8-digit addition or subtraction would be considerably more lengthy than in our example calculator, because of additional microinstructions for moving each digit to and from the main memory. (Our calculator handles such movement automatically, by means of its recirculating registers.) But the microprocessor chip after it became standardized would probably be *more economical* because it would be made in *larger quantities* for a much *wider variety of applications*.

To summarize what we've learned about microprocessors, we can say that they are a sort of *programmable building-block*, for use either singly or in groups. They can be used to make small computers or calculators or they can be programmed to serve in many other kinds of systems. Thus, they provide the opportunity to use a high functional density chip for many uses and get the cost advantages of high-volume production.

How do computers compare to the general programmed system picture?

Finally, let's say a few words about how "computers" fit the general picture of "general-purpose" programmed systems in *Figure 9-9*. Here, the sky's the limit, because the name "computer" applies to systems ranging from desk-top "microcomputers" to gargantuan number-crunchers that fill an entire room. A computer may be hidden inside a special-purpose system such as the autopilot we referred to in Chapter 5, or it may be a general-purpose data-processing facility that is available for running any kind of program the user needs.

The main difference between computers and microprocessors is the larger word size, greater memory capacity, increased capabilities and faster operation of the computer. However, the fundamental operations of both are the same.

Typically, however, we can say that a computer uses words of at least 16 bits (many use 32 bits, and some use 64 bits). Its ALU has a wider variety of functions to select from than a microprocessor's ALU. Some very fast, large computers have ALUs that even perform multiplication and division in a one-step, hard-wired fashion. And finally, most computers have a much larger main memory than microprocessors typically use—say, at least 65,536 words (2 to the sixteenth power), which is called a "65K-word" memory. Furthermore, outside the "main frame" (represented in *Figure 9-9*), there may be all manner of "peripheral" memory units, as we discussed with respect to *Figure 7-15*. And there may be any number of peripheral "input-output" units such as card readers and punching units, CRT terminals with keyboards and "television" screens, and so forth. This will become more clear to you as we discuss applications of computers in the next chapter.

However, regardless of all these complications, the main frame of any computer works very much as we have discussed with respect to *Figure 9-9*. In this respect, it is very similar to the example calculator with which you have now become familiar. Therefore, you should have a pretty good general idea of how a computer is put together from the building-blocks we learned about in Chapters 3, 4, and 7, and according to principles we discussed in Chapter 8. As a result, we have now seen the unifying principles of all digital systems, at work in the simplest and most complicated applications.

What's next?

And speaking of applications, that's the next main area for our discussion. Our learning so far has been pretty heavily oriented toward *how* digital systems work, without looking very hard at *where* they work. Now it's time to broaden our view and see how the things we have learned are *applied* in a *variety* of systems. That will be our main topic for the next chapter.

Quiz for Chapter 9

1. As integrated circuits become more complex, how are they prevented from becoming too specialized for economical high-volume production?
   a. It can't be done. This is the natural limit to the evolution of semiconductor circuitry.
   b. Certain hard-wired chips provide the circuitry for nearly any kind of system.
   c. One chip provides a wide selection of parts, and the user connects the parts to suit his application.
   d. By designing circuitry that can be programmed to suit a wide variety of applications.

2. What does "algorithm" mean?
   a. Analyzing a problem by breaking it into parts.
   b. A plan for breaking up problem solutions into steps.
   c. A group of stored instructions.
   d. A chart showing instructions as blocks.

3. Which "hard-wired" operations in our calculator form the basic steps used in all calculations?
   a. Add, subtract, multiply, divide.
   b. Add, subtract (or compare), and shift.
   c. AND, OR, NOT, NAND.
   d. Sense, decide, store, act.

4. What happens in the example calculator during the execution of a "conditional jump" instruction?
   a. The controller looks at the bit in the condition latch.

   b. A new address, provided in the current instruction, may be put into the address register.
   c. The stored numbers and flags are not changed.
   d. All of the above.

5. Which kind of instruction provides the main decision-making capability in a program?
   a. Register operations (add, subtract, compare, shift, relocate).
   b. Flag operations (read or write).
   c. Conditional jumps.
   d. Unconditional jumps.

6. Any program (or microprogram, routine, etc.) should be efficient in terms of:
   a. The number of instruction cycles required to execute it.
   b. The number of instructions it contains.
   c. How few of the system's parts it uses.
   d. A and B above.

7. What operations can typically be performed by an arithmetic and logic unit on incoming numbers or bits in the numbers?
   a. Addition and subtraction.
   b. Several kinds of comparisons.
   c. Several logic gating operations.
   d. Any of the above.

8. Which parts of the example
   calculator correspond to the
   working registers in a general
   programed system?
   a. The output stages of the
      number registers.
   b. The input stages of the
      number registers.
   c. The flag register and its
      control unit.
   d. A and B above.

9. In a general programmed
   system, what typically causes a
   microprogrammed routine to
   begin?
   a. External inputs, such as
      from a keyboard.
   b. Another microprogrammed
      routine.
   c. It's repeated automatically.
   d. An instruction in the main
      program.

10. How many levels of control
    operate in a general
    programmed system, counting
    hard-wired control of some
    parts?
    a. One.
    b. Two.
    c. Three.
    d. Four.

11. Judging by the discussion of
    microprocessors, which parts in
    the general programmed
    system in *Figure 9-9* are the
    minimum a set of chips should
    have to be called a
    microprocessor?
    a. The CPU, maybe without a
       microprogram memory.
    b. All the parts.

   c. The ALU, maybe with some
      registers.
   d. All but the input and output
      units.

12. One of the main features that
    distinguish microprocessors
    from computers is:
    a. Microprocessors always
       have fewer of the parts
       shown in *Figure 9-9*.
    b. Computers are never fully
       integrated.
    c. Words are usually shorter in
       microprocessors.
    d. Words are usually longer in
       microprocessors.

13. Microprocessors can be applied
    to make:
    a. Calculators.
    b. Computers.
    c. Many other kinds of digital
       systems.
    d. All of the above.

14. *Figure 9-9* represents the main
    frame of a computer, provided
    that:
    a. The main memory is
       reasonably large (say, 65 k
       words).
    b. The words have at least 16
       bits.
    c. It can handle many varieties
       of peripheral equipment.
    d. All of the above.

(Answers in back of the book)

Digital Electronics Today and in the Future

As we noted at the beginning of Chapter 8, there are so many different kinds of digital systems that we can't hope to comprehend them all in just one book. That's why we have depended on just one system (a calculator) for most of our examples in studying digital electronics. This approach has worked very well, because the calculator contains examples of most of the kinds of parts found in digital systems in general, put together in ways that illustrate important design factors.

However, our understanding has now advanced to the point that we can zoom the picture out and look at *many* kinds of present-day digital systems. So the first part of this chapter will be a survey of the applications of digital technology, as of the late nineteen-seventies.

Of course, as we have emphasized throughout this book, digital technology is *changing* rapidly, mainly because more gates and more storage capacity are being squeezed into integrated circuits. Consequently, the applications of digital methods are growing by leaps and bounds. And so the last part of the chapter will be a survey of trends for the future in digital electronics.

What are the main categories of digital applications?

Digital system's applications can be categorized into three groups: general-purpose data processing, automatic control and communications control.

In Chapter 8, we simplified the picture of system *functions* by studying three main design factors. Similarly, in this chapter we will simplify the picture of system *applications* by defining three main *categories* of applications, as listed in *Figure 10-1*. There are any number of ways to categorize digital applications. In this particular breakdown, we are grouping together systems that typically have certain design features in common, based on similarities in the jobs the systems do.

**Figure 10-1.
Categories of Digital
System Applications**

> **1. GENERAL-PURPOSE DATA PROCESSING**
>   A. Traditional "Business" Data Processing
>   B. Technical Computations
>   C. Simulation and Automatic Designing
>   D. Specialized Data Processing (including calculators, entertainment systems, and personal computing)
>
> **2. AUTOMATIC CONTROL**
>   A. Data Logging
>   B. Process Control
>   C. Sequencing of Events
>
> **3. COMMUNICATIONS CONTROL**

Although these categories are rather loose ones, this is still a useful way to help make sense out of a bewildering variety of digital applications.

What are "general-purpose data-processing" systems?

General purpose data-processing computers are fairly large systems with peripheral units containing additional disk or tape mass units and several specialized input/output units.

Taking the categories and subcategories in the order listed in *Figure 10-1*, let's first consider "general-purpose data-processing" systems. Here, we're mainly talking about certain kinds of medium to large *computers*, which may be represented as in *Figure 10-2*.

Beside the "main frame" that all computers have (consisting of a main random-access memory and a CPU as we noted in *Figure 9-9*), we typically have one or more magnetic tape and disk units as we saw in *Figure 7-15*. Furthermore, we have a terminal with a keyboard such as the CRT type that we discussed in Chapter 7, for direct interaction between an operator and the program being executed. And finally, there are typically several input and output units such as a card-punching device, a card reader, and a high-speed line printer. In this last group, there may be some more specialized input or output units that we will come to as we go along.

What features of computer operation do we need to be aware of?

Features of general-purpose data processing computer systems that are important are: processing words of different lengths, sorting and rearranging data to fit a variety of formats and coupling outputs to different I/O's.

As we discuss computer applications in this category (and the others in *Figure 10-1*), keep in mind the general features of computer operation that we've already covered. The various input devices form data into words of the length the computer can handle, and these words pass one at a time into certain addresses in the main random-access memory. The CPU (following its program of stored instructions interpreted by its controller) takes in words from certain memory addresses for arithmetic or logical processing by its ALU. New words that result are stored at certain addresses.

What's more important for some business applications, the CPU can *sort and rearrange* words in the memory without changing them. It does this by taking a word, looking at it or comparing it with another word, and deciding (according to the program) on a new address in which to put the word. Regardless of the processing details, however, the resulting words, in a certain order, are then transferred from the memory to one of the output units (tape, cards, printer, etc.)

One important capability of most computers that we haven't mentioned before is *variable word length*. Typically, a computer whose "regular" word length is 32 bits is also able to address and process 16-bit and 8-bit words, with the word length specified in the instruction for each operation. Obviously, the instruction must tell the ALU what kind of word it's dealing with, so it can do the right things to the right bits of the word.

Figure 10-2.
A General-Purpose Data
Processing Computer
System

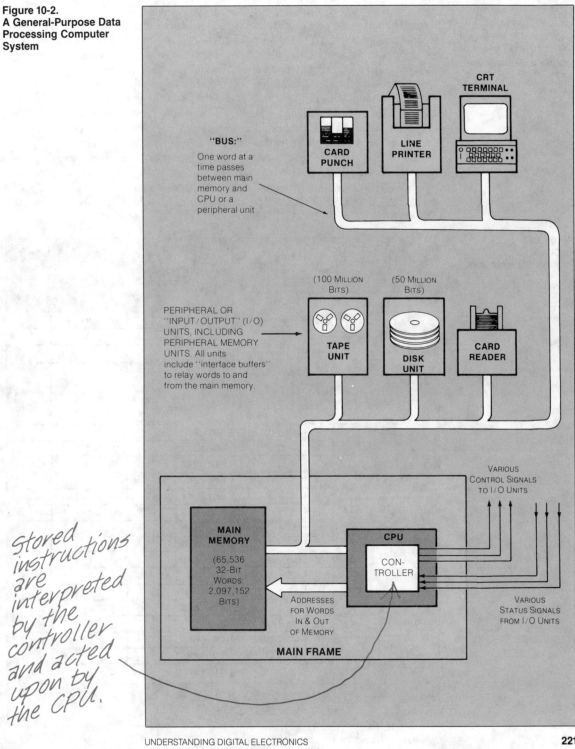

Stored
instructions
are
interpreted
by the
controller
and acted
upon by
the CPU.

Eight-bit words are used to represent letters, numerals, and symbols. The ASCII code is a common code used for this purpose.

The longer words are used for *numbers,* including sign (plus or minus) and decimal-point information. On the other hand, eight-bit words provide a method of representing *"characters" (letters, numerals, symbols) to be printed or displayed,* together with "control characters" that provide special signals to an output device (such as blank spaces, carriage returns, and so forth). *Figure 10-3* shows a popular code for this purpose, called "ASCII" (American Standard Code for Information Interchange; rhymes with "pass-key"). This code uses seven bits for each character, leaving the eighth bit of a word free for other handy purposes, such as a "word mark" to show that the character is the first computer word in a group that is read together. ASCII is especially useful in the communications field, which we will come to later.

With these general computer capabilities in mind, let's consider each of the subcategories of general-purpose data processing as listed in *Figure 10-1.*

**Figure 10-3.
ASCII - A
Communications Code**

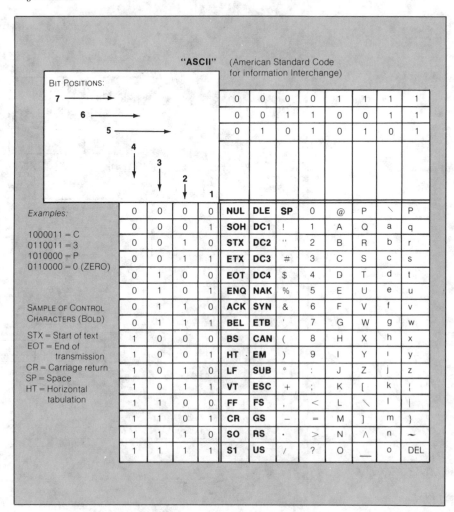

What's involved in "business" data processing?

"Business" data processing includes most of the present computer applications that affect the everyday life of the average person. It's the kind of processing used in *maintaining records* of great volumes of information, and in sorting, rearranging, and printing the information (or displaying it on a CRT screen) in useful ways. Typical applications include a business's records for accounting and bookkeeping, payroll, personnel data, inventory, and so forth.

For example *(Figure 10-4)*, your bank's daily records of transactions and balances for your checking account might be stored as numbers and characters in a short segment (perhaps a few inches) in a reel of magnetic tape. The rest of the reel would consist of similar records for other accounts, arranged in order according to account number.

Creating, maintaining, updating and printing banking and other business records is the primary function of business data processing. Thousands of transactions are handled quickly and accurately.

**Figure 10-4.
Processing a Bank's
Current Checking
Accounts**

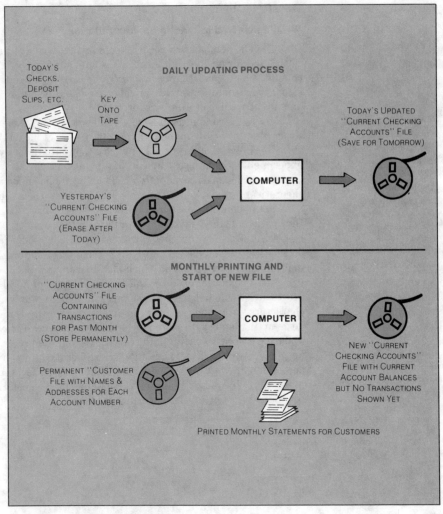

Once a day, as shown in *Figure 10-4* the "current checking account" reel would be loaded onto a computer tape unit, and another tape unit would be loaded with the tape containing the records of checks or deposits (or both) received for these accounts for that day. The computer would read the two tapes, enter the new transactions, figure the new balances, and record the updated accounts on a third reel of blank tape. Thus, a new current-checking-account reel is created every day for these particular account numbers.

Once a month, the information carried for these accounts is printed onto statements and mailed to customers. (Names and addresses are looked up in a permanent "customer file" tape.) The account balances are transferred to a fresh reel for use during the coming month, and the "old" reel is stored for a year or so as a permanent record of daily transactions for these accounts for that particular month. (The programs used for both the processes described would be stored in a disk unit and transferred to the main memory as needed.)

What's involved in "technical computations"?

The emphasis on record-keeping that we have just seen illustrated for business applications is not found in the "technical computations" subcategory in *Figure 10-1*. Here, we're talking about either long, complicated numerical calculations or numerical calculations on a great volume of numbers (or both).

Computers used in scientific research perform long, complex mathematical and scientific calculations, or a set group of calculations on a large volume of numbers.

For example, as illustrated in *Figure 10-5*, a scientific research experiment typically involves hundreds of sets of measurements from the experimental apparatus. For each set of data in many kinds of experiments, the two coordinates of a point on a chart must be calculated (a process called "data reduction"). Then a graph shaped according to a theoretical formula being tested is drawn so as to be as close as possible to as many points as possible (a process called "curve-fitting"). Finally, the average distance of all the points from the curve is calculated to see how well the theory matches the experimental evidence (a process called "statistical analysis"). Obviously, this technical computation job involves many calculations—some of which may be very complicated.

A computer used for this subcategory of general data processing would not need provisions for efficient handling of printed characters (provisions such as variable word length). But it would need a large main memory, to store all the data that typically needs to be looked at during the curve-fitting and statistical-analysis processes. And it would need a *plotter* output unit among its peripheral equipment, to plot data points and draw graphs. Furthermore, since many technical computation jobs require literally millions of very complicated calculations, the computer typically needs to work very fast.

**Figure 10-5.
Data Reduction and
Curve Fitting**

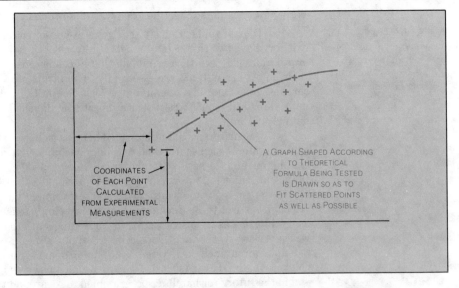

COORDINATES
OF EACH POINT
CALCULATED
FROM EXPERIMENTAL
MEASUREMENTS

A GRAPH SHAPED ACCORDING
TO THEORETICAL
FORMULA BEING TESTED
IS DRAWN SO AS TO
FIT SCATTERED POINTS
AS WELL AS POSSIBLE

What's involved in "simulation and automatic designing"?

Simulation of systems by computer is an important general data-processing application. Using computers, system performance can be predicted without actually building hardware.

The general data-processing category could be subdivided in many ways. But one other subcategory of particular interest is the use of a computer to *predict* how something being investigated or designed might operate (from an airplane in flight to a nation's gross national product), based on mathematical formulas developed to describe its operation. Formerly, *analog* computers were used for this purpose, using special electronic circuits in which varying voltages "simulated" (copied) the response of an object to a change in conditions. More and more, however, *digital* computers are used for such purposes (although the programming may be very laborious).

For example, a program may be written to calculate the slight bending and twisting of an airplane wing according to mathematical formulas involving the stiffness of the various parts, the weight of the engines hanging from the wing, and the speed and direction of the airplane's motion through the air. A "graphic display" terminal (a specially programmed CRT terminal) may be used to actually show a simplified drawing of the wing as it ponderously flutters and shakes. The position of each point and line in the drawing must be individually calculated at each interval in time. As you can imagine, a computer for this job must work extremely fast and have a very large main memory.

Furthermore, both technical-computation and simulation applications typically require that the computers use *very long words* (say, 64 bits or even more). This is so that the numerical fractions resulting from each calculation can be as *precise* as possible, to prevent errors from creeping in due to "rounding off" or "truncating" (cutting off) parts of each multiplication product or division quotient. Long words also permit very large numbers and very small fractions to be handled, as required by many computations in these two categories.

What other systems fall in the "general data-processing" category?

A specialized category under general-purpose data-processing systems would include video games, toys, teaching machines, etc.

Digital applications in the "general-purpose data-processing" category have two things in common. First, they typically use computers or programmed circuitry resembling that of computers. And second, their inputs and outputs are essentially to and from *people* rather than some other system. The remaining systems that fit these definitions are lumped together in *Figure 10-1* as a subcategory labelled "specialized data processing."

Calculators would be classified here, as well as certain very similar systems such as toys and teaching machines that resemble calculators, and the self-contained, stand-alone type of cash register that's not connected to a computer. This "catch-all" subcategory would also include most systems designed for entertainment purposes, although some of these (including games played on a television screen) use *simulation* as we discussed for the preceding subcategory. Here also, we could classify certain other specialized applications of computers. Examples might include a system that provides random access to large volumes of stored data (such as for card-catalog purposes in a library), and in the future for handling programmed tasks around the home. We'll say more about home applications later.

What's involved in the "automatic control" category?

Automatic control system applications use computers to control electrical or electrical-mechanical systems, receive inputs from sensors and output signals to actuators.

Moving on to the "automatic control" category of applications, we find a very wide variety of systems. Many of them formerly employed specially-designed hard-wired circuitry or analog methods but are now moving rapidly toward small computers and microprocessors instead. This category involves systems that are *directly hooked up* to external equipment or systems, for the purpose of *controlling* this equipment.

Automatic control systems have some or all of the parts represented in block-diagram form in *Figure 10-6*. The "system being controlled" may be anything from a soft-drink dispensing machine to a chemical reactor in a refinery. One or more streams of analog or digital information (or both kinds) flow into the digital system (bold outlines in *Figure 10-6*) by way of various sensing units and analog-to-digital converters. These streams provide information on the present status of the controlled system. Commands and reference information are provided by a human "operator" by way of switches, pushbuttons, and so forth.

Control signals developed by the digital system operate various "actuators" such as solenoids (electromagnets), motors, and the like, to produce the desired action in the controlled system. (Analog control signals require digital-to-analog converters as shown.) In addition, the digital system typically puts out some sort of "status" information, which may be lights, LED or CRT displays, printed reports, and so forth. The operator may use this information to exert direct manual control of the main system.

**Figure 10-6.
A Digital Control System**

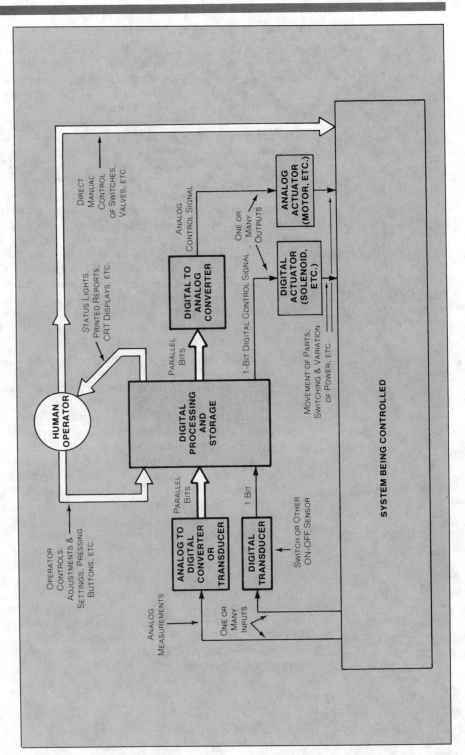

What are the subcategories of control systems?

The first of three general subcategories of control systems, data logging systems monitor and store information on a process for humans to base decisions on, rather than directly control the process.

"Data logging" systems have many inputs from the controlled system (up to hundreds of inputs), but do not exert automatic control. Instead, information is automatically collected and stored (that's what "logging" means) and presented to human operators, to guide their decisions on controlling the main system.

An example fitting neatly in this subcategory is a data-logger in an industrial plant that receives temperature signals from many points in a process. A small computer is programmed to "scan" all the temperature sensors (which are analog-to-digital measuring instruments), receiving data from one at a time on a rapidly repeated cycle. Temperatures are stored in a magnetic disk file, and are called up on command from the plant operator for output on an LED display. The software consists of permanently-stored microprogrammed routines, and the data words do not need to be very long. As with most systems containing a "dedicated" (built-in) computer, the operator is ordinarily not even aware that he's using a computer.

Process control computers initiate action based on the appropriate algorithm programmed into the system. Outputs are compared to human inputs (set points) to keep the system on track.

"Process control" systems contain small to medium-size computers programmed to not only scan a number of sensors but also to *compute appropriate control signals* to make the main system do what it's supposed to do. These computations follow algorithms based on knowledge about the processing being controlled.

For example, suppose the main system is a chemical reactor heated by steam coils. The operator puts in the desired temperature "set point" to be maintained. On a regularly repeated cycle, the computer determines how far off the actual temperature is. Then it calculates a number to control the steam valve, proportional to the temperature error (and perhaps also partly proportional to how fast the error is changing and how long the error has existed). Converted to an analog voltage, the output number regulates air pressure that drives an analog valve actuator to "steer" the temperature back to the desired set-point.

Specialized computers inside automatic control systems control the timing of the system.

A third subcategory of automatic control systems consists of units that do not calculate any *variable* outputs like process controllers but merely *switch things on and off*, usually several things in a prescribed sequence. Typically, one of the inputs to such a system is a *time signal* of some sort (which may be produced inside the system instead). Here, a good example is a traffic-light controller at a street intersection. In this case, electronic "transducers" connected to antenna loops imbedded in the pavement send pulses that tell the system (a microprocessor, nowadays) when there are cars waiting for a green light. (The microprocessor scans the transducers rapidly, looking for pulses.) Using different microprogrammed routines for various times of day as traffic patterns change, the microprocessor puts out switching signals to operate each light.

What's involved in "communications control"?

There's another large group of control applications that are so different from the scheme in *Figure 10-6* that they're set off to themselves as the last category in *Figure 10-1*. What's controlled here is the *routing* and *timing* of information sent from place to place, complete with details such as converting from one digital *code* to another, checking for *errors* in transmission, and *verifying* that messages were received. Somewhat the same techniques are involved whether the information is in *digital* form (say, words sent between a computer's main frame and a teletype terminal in the same room or another city) or *analog* form (such as telephone conversations).

The details of systems for communications control fall mostly in the field of computer architecture and operation, which is outside the scope of this book. However, to give you a general idea of the sort of application we're talking about, *Figure 10-7* shows in general terms the use of a computer for one important type of communications-control unit called a "peripheral and communications processor" (P&CP).

This subsystem, which contains a special type of microprocessor, acts as a sort of "master switchboard" for all information moving in and out of the main frame. It handles most of the petty, time-consuming details of shuttling words to and from the peripheral devices, leaving the CPU free for its main task of processing data inside the main frame.

What points about data communications are illustrated here?

Getting these characters to the typewriter terminal illustrates four important points that often appear in the communication of digital data:

First, the groups of bits are transmitted serially in a single line, "asynchronously," meaning without a common clock signal supplied to both ends of the line. This involves marking the beginning and end of each group with a zero for a "start" bit and two ones for "stop" bits. (See the example character "T" in *Figure 10-7*.)

Second, an extra bit called the "parity" bit is included at the end of each data group, before the "stop" bits. This bit tells whether the data group has an odd or even number of ones. It enables the receiving unit to find out whether a noise pulse has caused *any one of the bits* of a group to be received incorrectly.

Third, serial data is transmitted here in a *telephone line*, using certain tone pitches for ones and zeroes, by means of transducers (converters) called *"MODEMS"* (modulator-demodulators).

**Figure 10-7.
Using a Peripheral and
Control Processor**

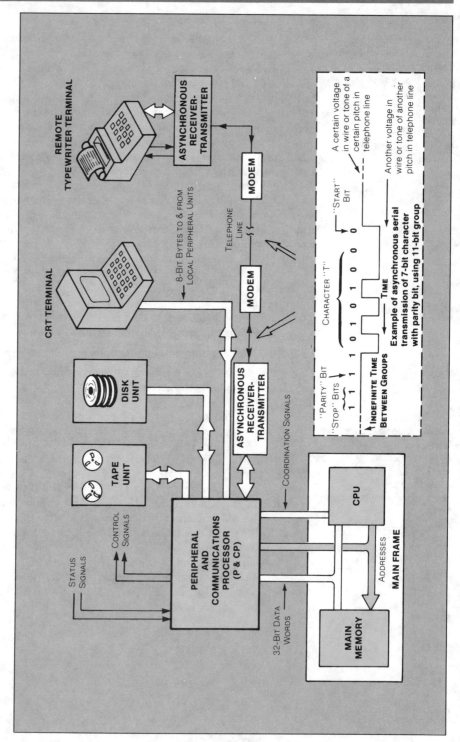

And fourth, transmitting data back and forth between independent systems (such as the main frame, the P&CP, and each peripheral unit) requires the use of definite rules called "protocol." This involves making sure everything is ready before sending one or more characters (words, bytes, etc.), and afterwards making sure the data was received properly.

You can see that communications control can be a very important and complicated matter, different from the first two application categories that we defined.

What have we learned about applications of digital electronics?

This survey of digital applications has not exactly been thorough enough to turn you into a system designer. But it should give you a general idea of the ways in which the basic principles you learned earlier are put to use for a variety of purposes. Keep in mind that many systems may fit into more than one category.

What's the general direction of progress in digital electronics?

As digital electronics continues to evolve, three areas of future progress/growth are becoming clear: increasing complexity, higher-density memory, and wider applications of programmed circuitry.

Turning now to look at the future, we can expect the most important changes to be continuations of trends we have already noted in three particular areas: increasing *complexity* of integrated circuits in general, new developments in high-density *memories*, and wider application of *programmed* circuitry. The resulting expansion of digital electronics will provide capabilities for data storage and processing, control, and communications in areas that were only dreamed of before. We are seeing this revolution in process now.

Let's consider each of the three areas of expected progress (complexity, memories, and programmed circuitry), beginning with complexity.

How can integrated circuits become more complex?

Making smaller devices and improving the accuracy of the diffusion and metal patterns by using improved photomasking technology, such as x-rays or electron beams, will yield more complex circuits.

Logic gates and bit-storage cells have already become so tiny and so closely-packed in integrated circuits that you may wonder how integrated circuits can possibly become *more* complex. But the fact is that we have a long way to go before we reach the theoretical limits of complexity for digital circuits formed at the surface of a semiconductor chip. Progress is mainly a matter of working the bugs out of various advanced manufacturing techniques. These techniques are becoming more and more sophisticated and expensive to work out and put into production. But the resulting ICs are expected to sell in quantities that increase fast enough to support this further development.

There are two main areas of research and development toward greater circuit complexity. The first pertains to methods for creating *smaller* diffused regions and metal patterns, and improving the *accuracy* with which the various diffusion patterns and metal patterns line up with one another on the same slice.

To create smaller patterns, *electron beams or X-rays* will have to be used in the place of ultraviolet light (as in *Figure 6-2c*) for hardening the photoresist. This is because electrons and X-rays have shorter wavelengths and thus create patterns with sharper, less fuzzy edges.

The second pertains to discovering and perfecting types of *circuits and devices that are smaller and simpler to process,* and that dissipate less heat while still switching rapidly. In this regard, there is still room for improvment in I²L circuitry, in similar kinds of inverted circuitry, and in the kinds and types of MOS devices and circuits.

Why are memory improvements particularly important?

Even greater impact will be made by developing low-cost, increased density memory units that can store significantly more bits and still maintain quick and easy access.

A second main ingredient of future progress in digital electronics is the development of memory units that store more bits in smaller space at less cost per bit, while providing quick access to any part of the stored information and transferring data in and out at high speed. As we mentioned in Chapter 7, such improvements are particularly important because so many bits of storage capacity are required in typical systems. Putting it another way, storing four times as many bits in the same space for the same cost will open up *far more new applications* than a similar reduction in the size and cost of logic gates.

There are two particularly promising new memory techniques that we should take a look at. Back in *Figure 7-14*, we mentioned two types of memory unit that fall midway between MOS shift registers and magnetic drums in the chart of cost per bit versus access time. These are new techniques called "charge-coupled devices" (CCDs) and "magnetic bubbles." Both involve chips similar to IC chips we have seen, and both provide *serial-access recirculating storage* of more bits than the IC methods we studied in Chapter 7, at a much lower cost per bit.

How do "charge-coupled device" memories work?

One candidate for improving high-density memory is a charge-coupled device (CCD). It can be viewed as a very simple, compact dynamic shift register in which positive charge buckets representing a 1 are pulsed along a row of metal spots timed by three clock phases. The absence of a charge is a zero.

Figure 10-8 describes the operating principle of CCDs. Here, we see a cross-section through a semiconductor chip as each phase of a three-phase clock network is activated (switched from zero to minus 12 volts). Tiny metal spots in a long row over thin oxide on a solid n-type substrate are alternately connected to the three clock phases. When a metal spot is pulsed to a negative voltage, it is capable of *attracting positive charge* to the underside of the oxide layer beneath it, much as a magnet can hold iron filings to the underside of a sheet of paper beneath it. The electric field from the negatively-charged metal spot actually creates a sort of invisible electric "bucket" that can hold positive charge.

As each clock phase returns to zero volts, all its buckets disappear, "dumping" their charges. But at the same time, another bucket is formed next to each one that dumps, by the negative voltage of the next clock phase. In this manner, a positive charge injected into the first bucket on the left in *Figure 10-8* will be dumped from bucket to bucket along the row of metal spots as one clock phase follows another. By providing amplifiers for injecting and detecting these charges, and letting the presence or absence of charge signify ones and zeroes, we create a very simple and compact type of *dynamic shift register*.

**Figure 10-8.
A 3-stage Charge-
Coupled Device**

The buckets will follow these metal spots in the manner of a dynamic shift register.

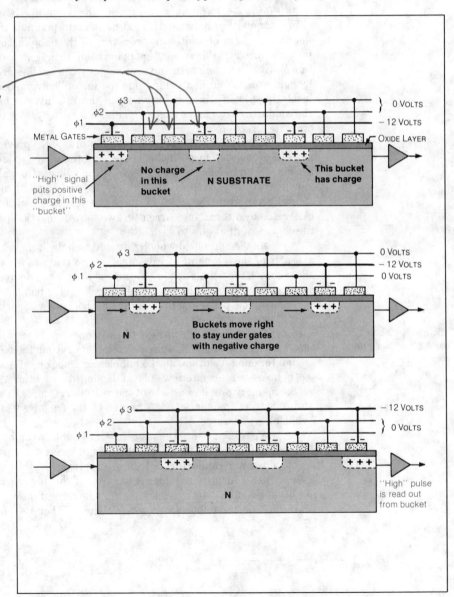

How does a magnetic bubble memory work?

Magnetic bubble memory is a second candidate for high-density memory. It uses tiny areas of magnetized regions, moving along a predetermined path in a special magnetic material inside a magnetic field, to perform the digital function.

Magnetic bubble memories produce much the same effect, although in a very different way. *Figure 10-9* shows what a magnetic bubble is. It's a tiny cylindrical magnetized region (about 0.2 mil or 5 micrometres in diameter) that can drift around in a thin film of certain magnetic crystalline materials such as "yttrium-iron garnet" (YIG). The YIG film is grown epitaxially (from a hot gas) on a slice of a non-magnetic crystalline material called "gadolinium-gallium garnet" or GGG. (The slice is originally cut from a larger crystal like a slice for an IC in *Figure 6-2*. Later, it is cut into memory chips like IC chips.) Each chip is sandwiched between two flat permanent magnets that provide a field pointing the opposite way from that of the bubbles, so that the bubbles form little "inverted" spots in the main magnetic field. (N and S in *Figure 10-9* mean north and south poles.) The bubbles are generated by pulses of current in an aluminum conductor "hairpin loop" over an insulating film of silicon oxide on top of the YIG layer. A bubble under the conductor loop can be destroyed in the same way, by a pulse in the opposite direction.

The bubbles are steered along definite paths provided by a pattern of magnetic iron-nickel strips over the silicon oxide film, which are not shown in *Figure 10-9*, but which you will see from above in the lower part of *Figure 10-10*. The bubbles are pushed along under these strips by a rotating "wobble" in the permanent magnetic field, generated by alternating current in two sets of coils that pass around the chip at right angles to each other.

As the main field wobbles around a circle, it makes little bar magnets of the segments of iron-nickel strips that are most nearly aligned with the field. Each bubble moves to seek the nearest north pole of a segment, and so the iron-nickel paths keep the bubbles spaced out—one bubble per segment along the path. (That's a spacing of about 0.8 mils or 22 micrometres).

The magnetic bubbles move sequentially under control of a magnetic field in a manner similar to a shift register. The presence or absence of a bubble provides the binary code.

The *presence of a bubble* stores a "one," and the *absence of a bubble* where one might be stores a "zero." Each time the field wobbles around for one revolution, all the bubbles advance one step. (As you can see, bubbles on alternate rows travel in opposite directions.) Thus, a bubble memory provides another type of simple, compact *shift register*, whose shifting frequency is determined by the frequency at which the "wobble" is generated.

We won't go into the ways in which bubbles can be switched to different paths, split in two, and detected (all by electric currents in simple aluminum conductors or iron-nickel strips like the ones in *Figures 10-9* and *10-10*.). Suffice it to say that this ingenious method, like other magnetic storage methods, has the advantage of *non-volatility*—the bubbles stay where they are when the power is turned off.

**Figure 10-9.
Magnetic Bubble
Memory**

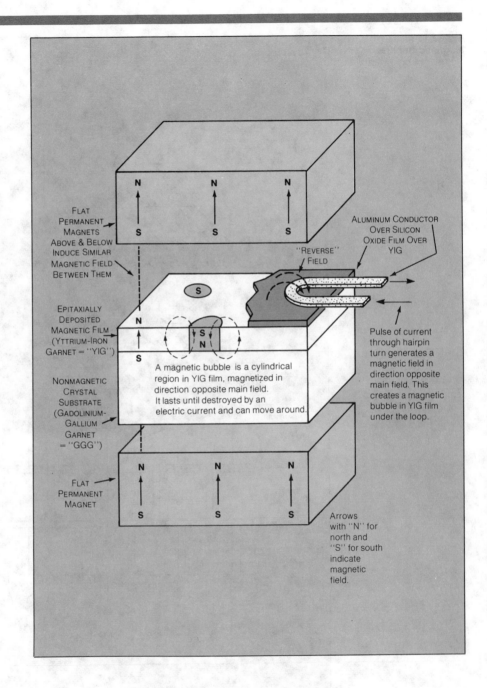

FLAT
PERMANENT
MAGNETS
ABOVE & BELOW
INDUCE SIMILAR
MAGNETIC FIELD
BETWEEN THEM

ALUMINUM CONDUCTOR
OVER SILICON
OXIDE FILM OVER
YIG

"REVERSE"
FIELD

EPITAXIALLY
DEPOSITED
MAGNETIC FILM
(YTTRIUM-IRON
GARNET = "YIG")

Pulse of current
through hairpin
turn generates a
magnetic field in
direction opposite
main field. This
creates a magnetic
bubble in YIG film
under the loop.

A magnetic bubble is a cylindrical
region in YIG film, magnetized in
direction opposite main field.
It lasts until destroyed by an
electric current and can move around.

NONMAGNETIC
CRYSTAL
SUBSTRATE
(GADOLINIUM-
GALLIUM
GARNET
= "GGG")

FLAT
PERMANENT
MAGNET

Arrows
with "N" for
north and
"S" for south
indicate
magnetic
field.

**Figure 10-10.
Magnetic Bubbles
Representing a Digital 1**

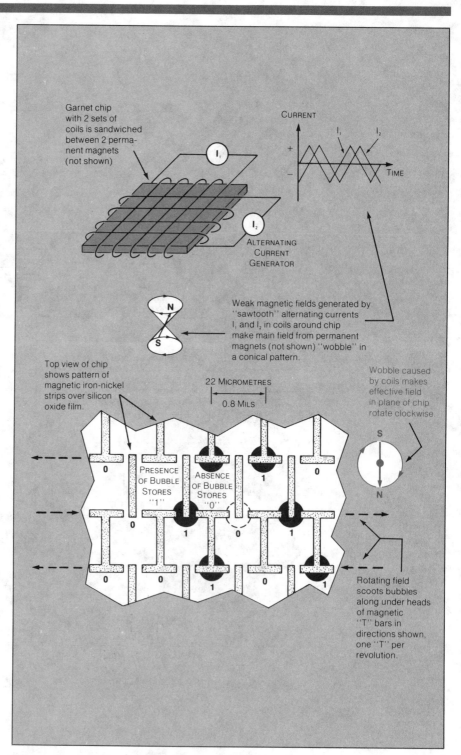

Garnet chip with 2 sets of coils is sandwiched between 2 permanent magnets (not shown)

I_1

CURRENT

I_1 I_2

+

−

TIME

I_2

ALTERNATING CURRENT GENERATOR

N

S

Weak magnetic fields generated by "sawtooth" alternating currents I_1 and I_2 in coils around chip make main field from permanent magnets (not shown) "wobble" in a conical pattern.

Top view of chip shows pattern of magnetic iron-nickel strips over silicon oxide film.

22 MICROMETRES

0.8 MILS

Wobble caused by coils makes effective field in plane of chip rotate clockwise.

S

N

PRESENCE OF BUBBLE STORES "1"

ABSENCE OF BUBBLE STORES "0"

0   0   1   0

0   1   0   1

0   0   1   0   1

Rotating field scoots bubbles along under heads of magnetic "T" bars in directions shown, one "T" per revolution.

What progress is expected in storage capacity of memory chips?

Memory units show great promise for tremendous increases due to advances in technology and manufacturing techniques.

Although nobody can predict how fast we will see improvements in cost and capacity of IC memory chips, *Figure 10-11* summarizes what we can expect for the near future with regard to capacity. The present numbers of bits per chip are as we saw in *Figure 7-14* (as of the middle to late nineteen seventies). Very soon, we will see these capacities *multiplied by two or four* as a result of developments now well under way.

Many persons expect that the most dramatic changes in digital systems will result from magnetic-bubble chips that could well hold a million or more bits in the not-too-distant future. Along with charge-coupled devices, these memories show promise of *replacing magnetic tape and disks* for small systems. To give you a point of reference, the entire text of this book could be stored in three or four one-million-bit chips, using 7-bit ASCII code *(Fig. 10-3)* for the characters.

**Figure 10-11.
Capacities of
Semiconductor Memory
Chips**

TYPE OF MEMORY	PRESENT	NEAR FUTURE
Static Ram	8k, 16k	32k
Dynamic Ram	16k, 64k	256k
Charge-Coupled Device	65k	128k
Magnetic Bubble	100k	256k
ROM and PROM	128k, 256k	512k

Why will progress involve programmed circuitry?

Programmable products with increased computing capability will be the trend of the future for digital computing systems.

Finally, the third area of progress for digital electronics that we pointed out earlier is in the use of *programmed* control rather than fully hard-wired circuitry. This trend (which was explained in Chapter 9 in terms of lower cost for unspecialized ICs that can be programmed for many applications) will become stronger as integrated circuits become more complex and more bits are stored in smaller chip areas, all at lower cost. The capabilities of integrated circuits will be especially extended by the storage of more and more elaborate programs as a result of larger and more economical memories.

One of the most remarkable results of this trend will be single IC chips having the information-processing power that a medium-sized *computer* had a few years ago. These ICs will be priced low enough to be used nearly as freely in systems as discrete transistors were used not too long ago.

We can only begin to imagine what kind of changes will result in digital electronics. Programmable integrated circuits will perform dedicated (built-in) functions that we're not even aware of in new data-processing, control, and communications applications. Chances are that each home will have at least one computer—perhaps connected to a public data-communications system and in communication with programmed circuitry in equipment such as intercoms, doorbells, air conditioners, garage-door openers, washing machines, typewriters, burglar alarms, television sets, telephones, and who knows what else? Considering the great volume of storage that will be available, such an integrated electronic network would go far beyond home computer tasks we can now envision, such as bookkeeping, storing recipes, etc. The possibilities are truly mind-boggling.

Where do you go from here?

Now that we've surveyed the various types of digital system applications and looked at the direction of progress in this field, we can say that we've finished the job of understanding digital electronics.

Although you may not feel confident enough to design your own system yet, you can probably understand the general operation of nearly any digital system you may read about. And more than that, you can probably grasp the *reasons* for various features of that design. This being so, you're not very far from creating your own systems, because much of the other information you need is a matter of reading the specifications of integrated circuits in manufacturer's data sheets and catalogs.

Where you go from here is up to you. Many other books are available for your further learning from Texas Instruments Learning Center, others may be found in libraries as well as bookstores and electronic supply stores. In addition, courses in this subject are available in many schools.

However, regardless of whether you pursue this subject further or move on to something else, you can be satisfied with having gained insight into the technology that, more than any other in history, is extending the mind and hands of man.

Quiz for Chapter 10

1. What do digital applications in the "general-purpose data-processing" category have in common?
   a. Use of computers or similar programmed circuitry.
   b. Their inputs and outputs are essentially to and from people.
   c. They control other systems.
   d. A and B above.

2. ASCII uses seven-bit groups to represent:
   a. Characters to be printed or displayed.
   b. Special signals to an output device.
   c. Numbers in a form for calculations.
   d. A and B above.

3. A computer with several magnetic tape units, a high-speed line printer, and variable word length would be particularly useful for:
   a. Traditional "business" data-processing.
   b. Technical computations.
   c. Simulation and automatic designing.
   d. Specialized data processing.

4. Computers used for technical computations or simulation and automatic designing typically need to have:
   a. Large main memory.
   b. High processing speed.
   c. Very long words.
   d. All of the above.

5. In which subcategory of Figure 10-1 would you classify a system that controls a coin-operated vending machine? (Inputs from coin-detection switches; outputs to solenoids to dispense change and products)
   a. Technical computations.
   b. Process control.
   c. Sequencing of events.
   d. Specialized data processing.

6. How does information on the status of the system being controlled flow into an "automatic control system"?
   a. The various streams of information flow simultaneously and automatically into the control system's memory.
   b. There's a separate little control system for each incoming stream of information.
   c. The control system "scans" the various inputs, receiving data from one at a time and deciding what to do with it.
   d. A and B above.

7. What purpose does a "parity bit" serve in communicating a group of bits from one system or subsystem to another?
   a. It tells whether there is an odd or even number of ones in the group.
   b. It provides a "word mark" to show that the character is the first of a group read together.
   c. It enables the receiving unit to determine whether one bit has been received incorrectly, by looking for an odd or even number of ones in the data group.
   d. A and C above.

8. How are bits transmitted serially in a telephone line?
   a. Higher voltage for 1, lower for 0.
   b. Waves of greater amplitude for 1, less amplitude for 0.
   c. Tone of one pitch for 1, tone of another pitch for 0.
   d. Long tone for 1, short tone for 0.

9. How can integrated circuits become more complex (more gates and storage cells in less area)?
   a. Creating smaller, more precisely aligned patterns on photoresist films, using electron beams and X-rays instead of light.
   b. Using types of circuits and devices that are smaller, simpler, etc.

   c. Working the bugs out of existing manufacturing techniques.
   d. A and B above.

10. More new applications will be opened up by improvements in the density and cost of:
    a. Logic circuitry (gates).
    b. Memory units.
    c. Input devices.
    d. Output devices.

11. Charge-coupled devices and magnetic bubbles provide density and cost improvements in:
    a. Random-access memory.
    b. Read-only memory.
    c. Serial-access recirculating memory.
    d. All of the above.

12. Non-volatility is an important advantage of:
    a. CCDs.
    b. Magnetic bubbles.
    c. Magnetic tape and disks.
    d. B and C above.

13. Which kind of skill would you say will be more important in understanding and working with digital electronics in the future?
    a. Analyzing and designing special electronic circuits.
    b. Analyzing and designing complete electronic systems.
    c. Selecting and designing software for programmable systems.
    d. A and B above.

(Answers in back of the book)

Glossary

Access Time: The average time it takes for a bit or word to be read at random in a memory unit. In a serial-access unit, this is half the time to go from one end of the stored data to the other.

Addend: A number added to another. (See "Augend.")

Adder: A building block capable of providing a sum and a carry, if required, when adding two numbers electronically.

Address: A binary number designating a particular location where information may be stored in a memory unit.

Algorithm: A plan for performing a job as a series of smaller jobs, typically speaking of a numerical calculation. More generally, a set of procedures by which a given result is obtained.

ALU: Arithmetic and Logic Unit. A subsystem that can perform any of a number of arithmetic and logical operations on words sent to it, such as addition, subtraction, comparison, and AND and OR functions of certain bits.

Amplitude: The height or strength of waves or pulses in an electric circuit or other transmission medium such as radio.

Analog: Electric analog *information* is information represented by a variable property of electricity, such as voltage, current, amplitude of waves or pulses, or frequency of waves or pulses. Analog *circuitry*, also called "linear" circuitry, is circuitry that varies certain properties of electricity continuously and smoothly over a certain range, rather than switching suddenly between certain levels.

AND Gate: A device or circuit with two or more inputs of binary digital information and one output, whose output is 1 only when all the inputs are 1. The output is 0 when any one or more inputs are 0.

Array: A group of elements arranged in a pattern.

ASCII: A code representing letters, numerals, punctuation marks, and control signals as seven-bit groups. It is used as a standard code by the U.S. for the transmission of data. The letters are an abbreviation for United States of American Standard Code for Information Interchange.

Asynchronous: Refers to circuitry and operations without common clock signals.

Augend: A number to which another is added. (See "Addend.")

BCD: BCD *Code:* See "Binary-Coded-Decimal Code." BCD *Counter:* See "Decade Counter."

Binary-Coded-Decimal (BCD) Code: A binary numbering system for representing each digit of a decimal number in groups of four bits. The binary value of these four bit groups is from 0000 to 1001 which code the decimal digits from "0" through "9".

Binary Number System or Code: A method of writing numbers by using two numeral digits, 0 and 1. Each position for a bit in a binary number represents 1,2,4,8,16,32, and so forth. Examples: 1000 = eight, 10101 = twenty-one.

Bipolar: A bipolar *transistor* is an n-p-n or a p-n-p transistor, as opposed to an MOS transistor. In a bipolar transistor, a small current through the "base" terminal controls a proportionately larger current through the "emitter" and "collector" terminals. A bipolar *integrated circuit* is one using bipolar transistors rather than MOS transistors.

Bit: The smallest possible piece of information. The simplest statement that can be made. A specification of one out of two possible alternatives. Usually thought of as a statement of yes or no. Bits are written as 1 for "yes" and 0 for "no". These symbols are the same as those used in binary numbers, so a bit is also a binary digit. (See "Binary Number System.")

Borrow: When a minuend digit is smaller than a subtrahend digit, 10 is added to the minuend digit and 1 is subtracted from the next more significant minuend digit. This procedure is called "borrowing" from the next more significant digit. The "borrowed" amount (1 or 0) is called the "borrow bit" or simply the "borrow." (See "Carry.")

Boolean Algebra: A simple system for writing combinations of logical statements (statements that can be either true or false). See *Figure 3-5.*

Branch: A program is said to "branch" where either of two possible instructions may be selected for execution next.

Buffer: A digital circuit with one input and one output. The output state is the same as that of the input. Used to "strengthen" a "weak" signal.

Building Block: In the field of electronic systems design, a combination of circuits or subsystems that can be used with other building blocks in many different combinations without redesigning. The simplest digital building blocks are logic gates.

Bus: Two or more conductors running in parallel used for carrying information. (Sometimes commonly used for power distribution conductors.)

Byte: A group of adjacent bits, usually a group formed for convenience in transmitting and receiving data (such as to and from a magnetic tape unit). Usually it takes more than one byte to make a word.

Calculator: A machine specially designed to perform arithmetic and computations which require numerous entries by the person operating the machine.

Capacitor: A device that stores electric charge in an electric or electronic circuit.

Carry: When the sum of two digits is equal to or greater than 10, then 10 is subtracted from this sum, and 1 is added to the next more significant digit of the sum. This procedure is called "carrying" to the next more significant digit. The "carried" amount (1 or 0) is called the "carry bit" or simply the "carry." (See "Borrow.")

CCD: Charge-coupled device. A means for very dense serial-access storage of bits as tiny packets of electric charge moving along the surface of a semiconductor chip.

Channel: In MOS transistors: the conducting portion between the main terminals, formed by the gate signal. In electronic systems: a path along which signals can be sent, such as data channel.

Character: A symbol whose image is formed by a display system for representation of information. Examples are numerals, letters, decimal point, punctuation marks, and special symbols indicating status of an electronic system. Examples are overflow or errors in a calculator.

Circuit: Strictly speaking, a circuit means a path for electricity from one terminal of a power supply (imagine a battery) through wires and usually through electrical or electronic devices, and back to the other terminal of the power supply. "Circuit" is also used to refer to a group of electric or electronic devices connected together to perform a certain job or function.

Clear: (See "Erase") To remove data and return all circuitry to an initial condition, usually "0"

Clock Input: An input terminal on a building-block typically used for receiving a timing control-clock signal, but used in some applications for a control signal or even data.

Clock Generator: A building block that generates clock signals.

Clock Signal: A regularly repeated signal supplied to more than one part of a system or subsystem to make things happen at the same time.

CMOS Integrated Circuit: A digital integrated circuit whose gates use both n-channel and p-channel MOS transistors in such a fashion that almost no current flows when a gate is not switching from one output state to another.

Code: A set of meanings or rules assigned to groups of bits. Each combination of bits (that is, each binary number that can be formed) has a certain meaning following certain rules in terms of a number, a character, an instruction, etc.

Code Converter: A name for a class of combinational building blocks that receive information in one code and transmit the same information in another code. With respect to one particular code (say, the BCD code used for numbers in a calculator), a building block that converts *to* this code is called an "encoder," and a building block that converts *from* this code is called a "decoder".

Combinational Building-Block or Network: A group of logic gates (perhaps just one) with no ability to store information, typically with several inputs and several outputs. For every combination of bits at the inputs, there is a definite, prearranged combination of bits at the outputs.

Comparator: A building-block that compares two binary numbers. There are several kinds of comparison: telling when the numbers are equal, when one is greater, when one is greater-than-or-equal-to the other, and so forth.

Complement: Usually means the "ones complement" of a bit, which is simply the *inverse* of the bit. To "complement" a number means to subtract it from a certain number (from one, in the case of ones complement).

Computer: A digital computer consists of at least one main frame, together with various peripheral input, output, and memory units. Distinguished from "minicomputer" and "microcomputer" by size and speed: longer words, larger memory, faster operation, more ALU operations, and generally more sophistication and flexibility.

Conductor: Something that conducts electricity from one place to another. It may be a wire, a metal strip on a printed-wiring board, a metal strip on the surface of an integrated-circuit chip, or a channel of semiconductor material inside an IC chip.

Controller: The parts (perhaps a subsystem) of a programmed system that select stored instructions, interpret them, and transmit control signals to the other parts of the system.

Core: A tiny ring of magnetic material that stores a bit as a permanent magnetic field. A "core memory" typically contains thousands or millions of cores.

Counter: A special kind of register made up of flip-flop circuits with one input and usually with a parallel output from each flip-flop, which counts pulses arriving at the input and stores the total count in a certain code (usually binary numbers).

CPU: Central Processing Unit. A section of a computer (or computer-like system) consisting of a controller, some registers, and an ALU.

CRT Terminal: Cathode-ray-tube terminal. A computer terminal with a screen similar to that of a television receiver, together with a keyboard.

D Flip-Flop: A clocked flip-flop with one data input (called "D"), whose "true" output changes at a clock signal to the state maintained at D during the clock signal.

Data: Another name for information. It may imply information being *transmitted* from one place to another, or information being *processed* as opposed to information used for *controlling* the processing of the other information.

Data Selector: A combinational building-block that routes data from one of several inputs to a single output, according to control signals. Also called "multiplexer." Two or more such one-bit selectors operating in parallel would be called a "two-bit data selector," etc.

Decade Counter: A modulo-ten counter, counting from zero to nine in BCD code.

Decimal Number System or Code: Also called "Arabic" number system. A method of writing numbers by using ten numeral digits. The "decimal digits" are 0, 1, 2, 3, 4, 5, 6, 7, 8, and 9. Each position for a digit in a decimal number stands for a place value of 1, 10, 100, 1000, and so forth.

Decoder: See "Code Converter." Loosely speaking, "decoder" may be the name used for a combinational building-block receiving several parallel inputs, which "recognizes" one or more combinations of input bits and puts out a signal when these combinations are received.

Dedicated: A dedicated system (say, a computer) is one that's limited to one particular job by the way it's built into a larger system (say, a process controller).

Demultiplexer: A combinational building-block that routes data from one input to one of several outputs, according to control signals.

Digit or Numeral Digit: One of the numerals or symbols used in a number system.

Digital: Symbolic of data in the form of pieces, i.e., bits or digits.

Diode: Any electronic device with two electrodes (terminals). Usually means a "p-n junction diode," which is a rectifying device (passes current only one way) consisting of a p region and an n region touching each other in the same semiconductor crystal.

Dividend: A number divided by another. (See "Divisor.")

Divisor: A number divided into another. (See "Dividend.")

Dopant, Doping: A substance added to semiconductor material to make it p-type or n-type is called a "dopant". Adding this substance is called "doping".

Driver: A circuit that provides digital signals with enough power to "drive" (operate) something other than a few nearby gates, such as LEDS in a display, magnetic cores in a memory, etc.

Dynamic Storage Circuit (or "Unit"): An electric circuit that stores one bit in the form of an electric charge in a capacitor (or in part of a circuit that acts as a capacitor). Two such circuits may be connected together to store a bit in master-slave style.

ECL: Emitter-coupled logic. Also, called "current-mode logic." General design principle for a bipolar logic gate that achieves fast propagation delay by diverting internal current from one path to another rather than switching saturated transistors on and off. A family of integrated circuits using gates like this.

Electric: Something that uses electricity. Some electric devices and circuits and systems are also "electronic."

Electronic: Something that uses electronic devices. Electronic devices are "vacuum" tubes (including gas-filled electron tubes) and solid-state semiconductor devices.

Encoder: See "Code Converter."

Epitaxial: An "epitaxial layer" on a semiconductor slice or chip is a thin layer of semiconductor crystal deposited on a substrate by crystal growth from a hot gas.

EPROM: Eraseable and programmable read-only memory. An IC memory chip whose stored data can be read at random. The data can be erased and new data can be stored, but only by a special system other than the one in which the memory is used.

Exclusive-Or Gate: A device or circuit with two (not more) inputs of binary digital information and one output, whose output is 1 when either input is 1 and 0 if neither or both inputs are 1. It acts as a certain combination of other gates.

Family: A family of digital integrated circuits is a group of ICs that use the same general design style for all gates, are processed during manufacture in much the same way, and whose input and output signals are all "compatible" with one another so that one can transmit to another.

Fan-Out: See Page 121.

Flag: A flag is a bit stored in a certain place, which the system uses as a "reminder" of something that's been done or something that needs to be done.

Flip-Flop: A building-block having two stable states that stores one bit by means of two gates (ordinarily NAND or NOR gates) "cross-coupled" as a latch, with the output of each forming an input to the other. It is capable of changing from one state to the other by the application of a control signal, but will also remain in that state after removal of the signal. A master-slave flip-flop contains two such latches. Most flip-flops contain additional features to make them more versatile.

Frequency: How often regular waves or pulses occur in a circuit or other transmission medium such as radio. Frequency is measured in hertz (cycles per second) and multiples of hertz.

Function Table: The function table for an electric or electronic binary digital circuit shows, the output electrical state that results from each combination of electrical states at the inputs. For binary electronic circuits, the states are either "high" or "low", "on" or "off", "open" or "closed." In examples in this book where voltages are not specified, "on" means high, and "off" means low.

Gate: A "logic gate" is an AND, OR, NOT, NAND, NOR or "Exclusive-OR" gate. In an MOS transistor, the "gate" is the metal plate for holding the charge to control the transistor.

Hardware: The actual physical parts and structure of a system, subsystem, etc., as opposed to "software" that may control its operation.

Hard-Wired: Describes a system, subsystem or building-block which does not contain stored instructions that control its operation. Its operation depends only on the way it is put together, and on inputs it receives.

IC: See "Integrated Circuit,"

I²L: I-squared-L, or "integrated injection logic." A certain type of logic gate that uses essentially only one bipolar transistor. A family of integrated circuits based on gates like this.

Inductor: Any device that makes electricity interact with a magnetic field, typically by means of a coil of wire.

Input: An information signal coming into a system or a part of a system. Can also mean the wire that carries this incoming information.

Instruction: A string of bits in a certain combination, stored in a "programmed" digital system such as a calculator. Each instruction contains information in a special code that tells the system what to do next. (See "microprogram.")

Instruction Cycle: The period of time during which a programmed system obeys a particular instruction.

Integer: A "whole number" not containing a fraction, decimal point, etc.

Integrated Circuit ("IC"): A small package with electrical terminals, containing a chip of silicon. The surface of the silicon is processed to form hundreds or thousands of transistors and other devices. These make up an electronic circuit.

Inverter: A binary digital building-block with one input and one output. The output state is the inverse (opposite) of the input state.

J-K Flip-Flop: A clocked flip-flop with two inputs called J and K which acts as a clocked R-S flip-flop, except that when clocked while both J and K are 1, it toggles to the opposite state rather than an unknown state like an R-S flip-flop.

Latch: A one-bit latch is a circuit with an input and an output for digital information, and a third input for a control signal. The control signal makes the output either follow the input or be held in its present state. Several latches used together with the same control signal to each (say four of them) would be a type of *register* called a 4-bit latch. Also, see "R-S" Latch" and "S-R Latch."

LED: See "Light-emitting diode."

Light-Emitting Diode (LED): A sort of semiconductor "light bulb" made of a small piece of semiconductor material (such as gallium phosphide) that makes light when electric current is passed through it in a particular direction, by way of two terminals.

Linear Circuitry: See "Analog."

Logic Diagram: A diagram using symbols for gates, flip-flops, building-blocks, etc., and showing connections between these parts.

Logic Gate: See "Gate."

LSB or LSD: Least-significant bit or digit. The bit or digit at the "right" end of a number, with the smallest numerical value.

LSI: A level of complexity of integrated circuits, such that a complete major subsystem containing 100 or more gates (or equivalent circuits) is fabricated in one integrated circuit.

Magnetic Bubble: A tiny moveable magnetized region formed under certain conditions in a thin film of magnetic garnet crystal fabricated similar to an IC. Such bubbles provide very dense serial-access storage of bits.

Magnetic Tape, Drum, Disk: Types of serial-access mass memory that store bits as tiny magnetized spots in moving magnetic material, used for peripheral storage in a computer system.

Main Frame: A section of a computer consisting of a CPU and a random-access mass memory.

Mask: See "Photomask."

Mass Memory: A memory that can store a relatively large amount of inforamtion and hold that information "permanently" until replaced by the system.

Master-Slave: A method of connecting two flip-flops or dynamic storage units to store one bit, so that the "master" unit receives and stores an incoming bit before the "slave" unit releases a bit that was previously received by the master unit. This arrangement permits bits to be transferred between flip-flops by using a common clock signal (or a common set of clock phases).

Memory: In a digital system, a "memory" or a "memory unit" is a part of the system where information is stored. (See "Mass Memory.")

Microaddress: See "Microprogram."

Microcomputer: A computer in the lowest range of size and speed, generally smaller, slower, and less sophisticated than a "minicomputer."

Microinstruction: See "Microprogram."

Microprocessor: An IC (or set of a few ICs) that can be programmed with stored instructions to perform a wide variety of functions, consisting at least of a controller, some registers, and some sort of ALU (that is, the basic parts of a simple CPU).

Microprogram: Certain programmed systems are "microprogrammed," meaning that they have two levels of programming. Each instruction in the "upper" or "main" level typically causes execution of a routine at the lower or "microprogrammed" level. Such a routine consists of "microinstructions" stored at "microaddresses" in a memory unit that's typically a ROM, PROM, or EPROM. Some systems (such as non-programmable calculators) operate only with microprogramming.

Minicomputer: A computer in a certain range of size and speed, generally smaller, slower, and less sophisticated than a "computer."

Minuend: A number from which another is subtracted. (See "Subtrahend.")

Modulus: The modulus of a counter is the number of states it counts through before returning to the beginning state. Written as "modulo" when used as a prefix, as in "modulo-12 counter."

Modulation: The controlling of a certain property of electricity so as to transmit information in analog form.

MOS Integrated Circuit: A digital integrated circuit whose transistors are all (or nearly all) MOS transistors. Varieties include n-channel MOS, p-channel MOS, and CMOS integrated circuits.

MOS Transistor: A class of transistors that operate by means of an electric field produced by a voltage on a metal plate called the "gate." The field acts through a thin layer of oxide insulation upon a semiconductor channel, controlling its depth and therefore controlling current through the channel.

MSB or MSD: Most-significant bit or digit. The bit or digit at the "left" end of a number, with the largest numerical value.

Multiplicand: A number multiplied by another. "Two times three" means three taken two times, so the multiplicand is three. (Two is the "multiplier.")

Multiplier: A number by which another is multiplied. (See "Multiplicand.")

Multiplexer: See "Data Selector."

N-Type Semiconductor Material: Semiconductor material (such as silicon) containing a very small proportion of certain other elements (such as phosphorus), causing any current through the material to be conducted mainly by the movement of *negative* charges ("free" electrons).

NAND Gate: A binary digital building-block that acts as an AND gate followed by an inverter.

Negative Logic: In electronic binary digital circuits, this means the decision to let the more negative of the two voltage levels represent 1 and to let the less negative level represent 0. *In this book, negative logic is not used unless it's specifically stated.*

Noise: Any signal that isn't supposed to be there. Electrical noise may be caused by small, irregular sparks when a switch is opened or closed. Or it may be caused by radio waves or by electric or magnetic fields generated by one wire and picked up by another.

Noise Margin: See page 121.

NOR Gate: A binary digital building-block that acts as an OR gate followed by an inverter.

NOT Gate: Occasionally used to mean "inverter."

Numeral or Numeral Digit: See "Digit."

OR Gate: A device or circuit with two or more inputs of binary digital information and one output, whose output is 1 when any one or more inputs are 1. The output is 0 only when all inputs are 0.

Output: An information signal going out of a system or a part of a system. Can also mean the wire that carries this outgoing information. (See also I/O.)

P-Type Semiconductor Material: Semiconductor material (such as silicon) containing a very small proportion of certain other elements (such as boron), causing any current through the material to be conducted mainly by the movement of *positive* charges ("holes" among "bound" electrons).

Parallel Circuit Connection: Two or more electrical devices (such as switches or transistors) are said to be connected "in parallel" when two terminals of each device are connected to the same two points, so that current can pass from one point to the other through any of the devices.

Parallel Data Transmission: Two or more bits of a group are said to be transmitted "in parallel" when they are all transmitted at the same time (as in a group of wires) from the same source to the same destination.

Parallel Register: Two or more flip-flops (or dynamic storage units) with a common clock signal, used to store bits transmitted in parallel.

Peripheral: In a computer system, "peripheral" units or equipment are those outside the main frame, including disk and tape units, printers, keyboard terminals, etc.

Phase: The time interval for each clock "cycle" in a system may be divided into two or more "phases" (like the phases of the moon). The phases are defined by pulses in a separate network of wires for each phase. During a particular phase, the signal in that clock network is in the state defined as "active" (let's say the "high" or "1" state). The clock cycles are repeated over and over again, phase by phase. The phases provide a method of making several things happen in the proper order during one clock cycle.

Photomask: A transparent glass plate carrying an intricate, very precise pattern of microscopically small opaque (dark) spots photographically reduced from a larger pattern. The opaque spots represent areas on a semiconductor slice into which a dopant will be diffused, or areas of metal that will be etched away.

Photoresist: A liquid that, when spread in a thin film and dried, quickly hardens into a tough plastic substance where struck by ultraviolet light. When the unhardened areas have been washed away, the material beneath is exposed for etching by an acid. The hardened areas "resist" the acid.

PLA: Programmable Logic Array. An MOS read-only memory used as a network of logic gates.

Positive Logic: In electronic binary digital circuits, this means the decision to let the "higher," more positive of the two voltage levels represent 1 and to let the "lower," more negative level represent 0. Positive logic is always assumed in this book unless "negative logic" is specifically stated.

Potentiometer: A variable resistor with three terminals, normally used for transmitting a manually variable voltage ("potential") part-way between voltages supplied at two main terminals of the device.

Power Dissipation: See page 120.

Product: In arithmetic, the result of a multiplication. In Boolean algebra, the AND function of two or more variables.

Program: In a programmed system, any group of instructions that are "programmed," meaning planned so that one instruction leads to another.

Program Counter: In a programmed system, the parts that provide a method for adding 1 to the address of the current instruction. May consist of the address register and an adder.

Programmed System: A system that operates by following a series of stored instructions. Also called "variable-program" system, since the instructions can be changed.

PROM: Programmable read-only memory. An IC memory chip whose stored data can be read at random. The data is stored permanently by the user after the chip is manufactured, and cannot be changed afterward. Storing data in this fashion is loosely called "programming" the memory.

Propagation Delay: See "Switching Speed," page 120.

Quotient: The result of a division.

RAM: Random-access memory. A memory with a number of storage locations, where words may be "written" (stored) or "read" (recovered) in any order at random.

Refresh: To refresh a dynamic storage unit means to restore its charge to the desired voltage level.

Register: A certain type of temporary storage unit for digital information. (See "Shift Register" and "Parallel Register.")

Relay or Electromechanical Relay: A mechanical switch with a moving contact called an "armature" that is moved by a magnetic field generated by electricity in a coil of wire.

Reset: To "reset" a stored bit means make it a "0."

ROM: Read-only memory. A memory unit containing data permanently stored when the unit was manufactured. Usually an IC chip with each bit stored as a permanent electrical connection of some sort, which can be read at random.

Routine or Programmed Routine: A series of instructions followed by a programmed system in doing a particular job. Usually contained within a main program. It may occur over and over again. The instructions are "programmed," meaning planned so that each instruction leads to another. A routine can even be one instruction followed over and over again.

R-S Flip-Flop: Any of several kinds of flip-flops, in which a momentary 1 at the R (reset) input changes the "true" output to 0, and a momentary 1 at the S (set) input changes the "true" output to 1. Varieties include clocking and master-slave features.

R-S Latch: A simple kind of R-S flip-flop. Two NOR gates cross-coupled, with the output of each forming an input to the other.

$\overline{\text{S}}$-$\overline{\text{R}}$ **(S-Bar, R-Bar) Latch:** A simple kind of flip-flop. Two NAND gates cross-coupled, with the output of each forming an input to the other.

Schottky Diode: A type of rectifying (one-way) diode formed simply by a metal terminal contacting a lightly-doped region in a semiconductor crystal. Its low "forward voltage drop" and quick response to pulses make it useful in improving the performance of TTL circuits.

Segment: In this book, "segment" means one of the seven bars in a rectangular figure-8 pattern in a "7-segment" character display.

Semiconductor: "Semiconductor material" is various solid substances such as silicon, germanium, and gallium phosphide. When made in nearly pure crystal form, these materials conduct electricity in very special, useful ways. A "semiconductor device" (or just a "semiconductor") is a transistor or diode (or other similar things) made using semiconductor material.

Sequential Building-Block or Network: One or more flip-flops or dynamic storage units, typically with one or more logic gates, and typically with several inputs and several outputs. The combination of bits at the outputs does not depend only on the combination at the inputs at the present moment, but on past history of a sequence of input combinations over a period of time.

Serial-Access Memory: A memory in which the stored data is accessible for reading or writing only in a definite, fixed order rather than at random.

Serial Data Transmission: Two or more bits of a group are said to be transmitted "in series" when one at a time is transmitted through the same wire. Such transmission is called "serial" transmission.

Series Circuit Connection: Two or more electrical devices (such as switches or transistors) are said to be connected "in series" when they form a chain from one point to another, so that the same current flows through all of them.

Set: To "set" a stored bit means make it a "1."

Shift: A movement of stored data right or left.

Shift Register: Two or more flip-flops (or dynamic storage units) with a common clock signal, connected in series so that stored bits shift one stage during each clock cycle.

Signal: A word used in describing the operation of electric or electronic circuits. It means electrical voltage or current or waves carrying information, or the information itself.

Software: The stored instructions (or "program") that control a programmed system, as opposed to the "hardware" of which the system is physically constructed.

Speed-Power Product: See Page 121.

State: The "logic state" of a conductor in a digital circuit means its condition as to whether it is carrying a 1 or 0.

Static Memory: An IC memory whose storage elements consist of flip-flops rather than dynamic storage units.

Stored Program: A set of instructions in memory determining the order of the problem solution.

Substrate: Literally, "underlayer." The semiconductor material of a slice or chip that lies beneath the diffused and epitaxially deposited regions.

Subroutine: A routine that is part of another routine. (See "Routine.")

Subsystem: A smaller system inside a larger system. Each subsystem can be thought of as a separate system with its own job to do.

Subtrahend: A number subtracted from another. (See "Minuend.")

Sum: In arithmetic, the result of an addition. In Boolean algebra, the OR function of two or more variables.

Switching Circuit: An electric or electronic circuit whose output (or outputs) are in definite electrical states, rather than varying over a wide range. The switching circuits used in binary digital systems have only two possible states. These are usually two different voltage levels.

Synchronous: Refers to two or more things made to happen in a system at the same time, by means of a common clock signal.

System: In general, a system is a group of things that work together as a unified whole. In electronics, a system is a group of devices or circuits or subsystems that work together to do a certain job.

T Flip-Flop: A flipflop with an input called "T," whose outputs "toggle" to the opposite states on receiving a signal at T.

Terminal: A computer terminal is an input and output device operated by a person. Input is usually by a keyboard. Output is usually by typewriter or CRT screen.

Transducer: A device that converts information from one medium (say, the position of a control knob) to another (say, electronic digital signals).

Truth Table: The truth table for a binary digital building-block shows, for each information output, the logic state that results from each combination of logic states at the information inputs. The logic states are 1 (yes, true) and 0 (no, false). (See "Function Table.")

TTL or T²L: Transistor-transistor logic. A certain style of circuit design for a logic gate using two bipolar transistors. A broad family of integrated circuit types whose gates employ the general principle of this basic two-transistor arrangement.

USASCII: See "ASCII."

Variable: A quantity that can assume any of a given set of values.

Volatile: A volatile memory is one that loses its stored data when the electric power is turned off.

Word: A group of bits or string of bits handled as a unit usually stored at a certain address in a random-access memory (RAM).

Write: To record data in a storage or data medium.

YIG: Yttrium-iron garnet. A magnetic crystalline material suitable for holding magnetic bubbles.

Index

UNDERSTANDING DIGITAL ELECTRONICS

Answers to Quizzes

Chapter 1
1. c
2. c
3. d
4. d
5. a
6. d
7. d
8. b
9. d
10. c
11. d
12. a
13. d
14. c

Chapter 2
1. d
2. a
3. c
4. a
5. d
6. b
7. a
8. b
9. d
10. b
11. d
12. a
13. d
14. a
15. b
16. a
17. c
18. d

Chapter 3
1. b
2. c
3. b
4. b
5. a
6. d
7. b
8. c
9. d

Chapter 4
1. b
2. c
3. a
4. b
5. b
6. a
7. a
8. c
9. b
10. d
11. b
12. c
13. d
14. d
15. c

Chapter 5
1. d
2. c
3. d
4. c
5. a
6. b
7. d
8. d
9. d
10. c
11. c
12. d
13. a

Chapter 6
1. a
2. b
3. c
4. c
5. d
6. c
7. c
8. d
9. d

Chapter 7
1. c
2. d
3. c
4. c
5. c
6. d
7. a
8. b
9. d
10. c
11. d
12. a
13. d

Chapter 8
1. b
2. a
3. b
4. d
5. d
6. b
7. b

Chapter 9
1. d
2. b
3. b
4. d
5. c
6. d
7. d
8. d
9. d
10. c
11. a
12. c
13. d
14. d

Chapter 10
1. d
2. d
3. a
4. d
5. c
6. c
7. d
8. c
9. d
10. b
11. c
12. d
13. c